T0304120

Global Logistics Management

Industrial Engineering: Management, Tools, and Applications

*Industrial Engineering Non-Traditional Applications
in International Settings*

Industrial Engineering Applications in Emerging Countries

Global Logistics Management

Global Logistics Management

Edited by
Bahar Y. Kara
İhsan Sabuncuoğlu
Bopaya Bidanda

CRC Press
Taylor & Francis Group
Boca Raton London New York

CRC Press is an imprint of the
Taylor & Francis Group, an **informa** business

CRC Press
Taylor & Francis Group
6000 Broken Sound Parkway NW, Suite 300
Boca Raton, FL 33487-2742

© 2015 by Taylor & Francis Group, LLC
CRC Press is an imprint of Taylor & Francis Group, an Informa business

No claim to original U.S. Government works

International Standard Book Number-13: 978-1-4822-2694-2 (Hardback)

Library of Congress Cataloging-in-Publication Data

Global logistics management / editors, Bahar Y. Kara, Ihsan Sabuncuoglu, and Bopaya Bidanda.
    pages cm
Includes bibliographical references and index.
ISBN 978-1-4822-2694-2 (alk. paper)
    1. Business logistics--Mathematical models. 2. Production engineering. I. Kara, Bahar Y. (Bahar Yetis) II. Sabuncuoglu Ihsan (Engineer) III. Bidanda, Bopaya.

HD38.5.G577 2014
658.5--dc23   2014023515

Visit the Taylor & Francis Web site at
http://www.taylorandfrancis.com

and the CRC Press Web site at
http://www.crcpress.com

This book is dedicated to my dear husband Kadri,
my lovely daughters Sıla and Maya, and my dear father İmdat,
who have always been the inspiration of my life.

Bahar Y. Kara

Contents

VIII   CONTENTS

Preface

We are pleased to present this book that focuses on a key concept of industrial engineering that has received increased attention throughout the years—*logistics*. Starting from classical location and routing models, the concept of logistics has evolved through different application areas in the last two decades. Especially with the development of the party logistics (PL) concept due to different levels of outsourcing logistic operations and with the increasing demand from industry, academic research interest in logistics has increased rapidly throughout the years. This book provides an outlet for recent developments in global logistics. It clearly illustrates logistics problems encountered in many different application areas and presents the state of the art of some classical applications.

We therefore believe that this book can create an awareness of the richness in the logistics applications of the industrial engineering discipline.

MATLAB® is a registered trademark of The MathWorks, Inc. For product information, please contact:

The MathWorks, Inc.
3 Apple Hill Drive
Natick, MA 01760-2098 USA
Tel: 508-647-7000
Fax: 508-647-7001
E-mail: info@mathworks.com
Web: www.mathworks.com

Editors

Bahar Y. Kara is an associate professor in the Department of Industrial Engineering at Bilkent University.

Dr. Kara earned an MS and a PhD from Bilkent University Industrial Engineering Department, and she worked as a postdoctoral researcher at McGill University in Canada.

Dr. Kara was awarded Research Excellence in PhD Studies by INFORMS (Institute for Operations Research and Management Science) UPS-SOLA.

In 2008, Dr. Kara was awarded the TUBA-GEBIP (National Young Researchers Career Development Grant) Award. She attended the World Economic Forum in China in 2009. For her research and projects, the IAP (Inter Academy Panel) and the TWAS (The Academy of Science for the Developing World) awarded her the IAPs Young Researchers Grant. Dr. Kara was elected as an associate member of the Turkish Academy of Sciences in 2012. She has been acting as a reviewer for the top research journals within her field. Her current research interests include distribution logistics, humanitarian logistics, hub location and hub network design, and hazardous material logistics.

İhsan Sabuncuoğlu is the founding rector of Abdullah Gul University. He earned his BS and MS in industrial engineering from the Middle East Technical University in 1982 and 1984, respectively. He earned his PhD in industrial engineering from Wichita State University in 1990.

Dr. Sabuncuoğlu worked for Boeing, Pizza Hut, and the National Institute of Heath in the United States during his PhD studies. He joined Bilkent University in 1990 and worked as a full-time faculty member until 2013. In the meantime, he held visiting positions at Carnegie Mellon University in the United States and at Institut Français de Mécanique Avancée (IFMA) in France. His research interests are in real-time scheduling, simulation optimization, and applications of quantitative methods to cancer-related health-care problems. His research has been funded by TUBITAK (The Scientific and Technological Research Council of Turkey) and EUREKA (a European-wide initiative to foster European competitiveness through cooperation among companies and research institutions in the field of advanced technologies).

Dr. Sabuncuoğlu also has significant industrial experience in aerospace, automotive, and military-based defense systems. His industrial projects are sponsored by a number of both national and international companies. He is currently the director of the Bilkent University Industry and the University Collaboration Center (USIM) and the chair of the Advanced Machinery and Manufacturing Group (MAKITEG) at TUBITAK.

In addition to publishing more than a hundred papers in international journals and conference proceedings, Dr. Sabuncuoğlu has edited two books. He is also on the editorial board of a number of scientific journals in the areas of industrial engineering and operations research. He is a member of the Institute of Industrial Engineering, the Institute for Operations Research, the Management Sciences, and the Simulation Society. He is also a member of the Council of Industrial Engineering Academic Department Heads (CIEADH) and various other professional and social committees.

 **Bopaya Bidanda** is currently the Ernest E. Roth professor and chairman in the Department of Industrial Engineering at the University of Pittsburgh. His research focuses on manufacturing systems, reverse engineering, product development, and project management. He has published five books and more than a hundred papers in international journals and conference proceedings. His recent (edited) books include those published by Springer— *Virtual Prototyping & Bio-Manufacturing in Medical Applications* and *Bio-Materials and Prototyping Applications in Medicine.* He has also given invited and keynote speeches in Asia, South America, Africa, and Europe. He also helped initiate and institutionalize the engineering program on the Semester at Sea voyage in 2004.

He previously served as the president of the Council of Industrial Engineering Academic Department Heads (CIEADH) and also on the Board of Trustees of the Institute of Industrial Engineers. He also serves on the international advisory boards of universities in India and South America.

Dr. Bidanda is a fellow of the Institute of Industrial Engineers and is currently a commissioner with the Engineering Accreditation Commission of ABET. In 2004, he was appointed a Fulbright Senior Specialist by the J. William Fulbright Foreign Scholarship Board and the U.S. Department of State. He received the 2012 John Imhoff Award for Global Excellence in Industrial Engineering given by the American Society for Engineering Education. He also received the International Federation of Engineering Education Societies (IFEES) 2012 Award for Global Excellence in Engineering Education in Buenos Aires and also the 2013 Albert G. Holzman Distinguished Educator Award given by the Institute of Industrial Engineers. In recognition of his services to the engineering discipline, the medical community, and the University of Pittsburgh, he was honored with the 2014 Chancellors Distinguished Public Service Award.

Contributors

Aysun Akış
Department of Industrial
  Engineering
Istanbul Kültür University
Istanbul, Turkey

Ayşe Nur Asaly
Department of Industrial
  Engineering
Koç University
Istanbul, Turkey

Mustafa Avci
Department of Industrial
  Engineering
Dokuz Eylül University
İzmir, Turkey

Emre Berk
Department of Management
Bilkent University
Ankara, Turkey

Hande Çakın
Department of Industrial
  Engineering
İzmir University of Economics
İzmir, Turkey

Aylin Çalışkan
Department of Industrial
  Engineering
İzmir University of Economics
İzmir, Turkey

Derya Dinler
Department of Industrial
  Engineering
Middle East Technical
  University
Ankara, Turkey

Mehtap Dursun
Department of Industrial
  Engineering
Galatasaray University
Istanbul, Turkey

E. Ghorbani-Totkaleh
Department of Industrial
    Engineering
Amirkabir University of
    Technology
Garmsar, Iran

Mahmut Ali Gökçe
Department of Industrial
    Engineering
İzmir University of Economics
İzmir, Turkey

Ülkü Gürler
Department of Industrial
    Engineering
Bilkent University
Ankara, Turkey

Ezel İlkyaz
Department of Industrial
    Engineering
İzmir University of Economics
İzmir, Turkey

Cem İyigün
Department of Industrial
    Engineering
Middle East Technical
    University
Ankara, Turkey

E. Ertugrul Karsak
Department of Industrial
    Engineering
Galatasaray University
Istanbul, Turkey

Ramez Kian
Department of Industrial
    Engineering
Bilkent University
Ankara, Turkey

Ezgi Kınacı
Department of Industrial
    Engineering
İzmir University of Economics
İzmir, Turkey

Mesut Kumru
Department of Industrial
    Engineering
Doğuş University
Istanbul, Turkey

Gürkan Mercan
Department of Industrial
    Engineering
İzmir University of Economics
İzmir, Turkey

M. Amin Nayeri
Department of Industrial
    Engineering and
    Management Systems
Amirkabir University of
    Technology
Tehran, Iran

Erdinç Öner
Department of Industrial
    Engineering
İzmir University of Economics
İzmir, Turkey

Deniz Ozdemir
Department of International
   Logistics Management
Yasar University
İzmir, Turkey

Syed Asif Raza
Department of Management
   and Marketing
College of Business and
   Economics
Qatar University
Doha, Qatar

Bekir Sahin
Department of Maritime
   Transportation and
   Management Engineering
Istanbul Technical University
Istanbul, Turkey
and
Surmene Faculty of Marine
   Sciences
Karadeniz Technical University
Trabzon, Turkey

M. Sheikh Sajadieh
Department of Industrial
   Engineering and
   Management Systems
Amirkabir University of
   Technology
Tehran, Iran

F. Sibel Salman
Department of Industrial
   Engineering
Koç University
Istanbul, Turkey

Zeynep Sener
Department of Industrial
   Engineering
Galatasaray University
Istanbul, Turkey

Yoshiaki Shimizu
Department of Mechanical
   Engineering
Toyohashi University of
   Technology
Toyohashi, Japan

Zeynel Sırma
Department of Industrial
   Engineering
Istanbul Kültür University
Istanbul, Turkey

Beril Sözer
Department of Industrial
   Engineering
İzmir University of Economics
İzmir, Turkey

Şeyda Topaloğlu
Department of Industrial
   Engineering
Dokuz Eylül University
İzmir, Turkey

Mustafa Kemal Tural
Department of Industrial
   Engineering
Middle East Technical
   University
Ankara, Turkey

Mihaela Turiac
Economic Cybernetics and
    Statistics Doctoral School
The Bucharest University of
    Economic Studies
Bucharest, Romania

Fadime Üney-Yüksektepe
Department of Industrial
    Engineering
Istanbul Kültür University
Istanbul, Turkey

Evrim Ursavas
Department of Operations
University of Groningen
Groningen, the Netherlands

Zeynep Yalçın
Department of Industrial
    Engineering
Istanbul Kültür University
Istanbul, Turkey

Introduction

Logistics is one of the key concepts of industrial engineering that has received increasing attention throughout the years. This book focuses on some recent developments in global logistics and will therefore provide some of the more exciting applications, developments, and implementations of classical operations research techniques on logistics problems. This book contains 13 chapters on topics ranging from continuous location models to disaster relief logistics. Chapters 1 and 2 are devoted to some variations and recent developments in the most classical logistic problem, the vehicle routing problem (VRP), followed by analyses and discussions on various logistics problems of airline and marine systems. The book details a wide range of application-oriented studies, ranging from a metropolitan bus routing problem to relief logistics. The problems encountered in continuous space deserve special attention and an overview of continuous location problems is provided in Chapter 11. Finally, Chapters 12 and 13 discuss the issue of consolidation, scheduling, and replenishment decisions together with routing.

In more detail, Chapter 1 focuses on a three-echelon logistics network and proposes a methodology that supports decision making at a tactical and operational level associated with daily inventory management. The methodology also includes a new method for solving multi-depot VRPs.

Especially with the growing interest in reverse logistics problems, the vehicle routing problem with simultaneous pick up and delivery (VRPSPD) has attracted more research interest. Chapter 2 presents a local search solution approach for this problem. The proposed approach is a hybrid version of simulated annealing and variable neighborhood descent methods and computationally outperformed the methods in the literature on benchmark instances.

Chapter 3 presents a novel approach for airline logistics including fare pricing and seat inventory control. Especially in today's competitive environment, market segmentation through differentiated pricing is a common practice and Chapter 3 provides an analysis for the airline industry.

Chapter 4 focuses on the berth–crane allocation problem in container terminals. Container terminal logistics became an active research area with the increasing trend in seaborne trade. The chapter proposes a decision support tool for simultaneous berth allocation and crane scheduling problems.

Another marine system logistics application is analyzed in Chapter 5 where Arctic transportation is being investigated. Especially due to global warming, the Arctic Ocean is more navigable and Chapter 5 investigates ice navigation problems and proposes a model stating factors that affect ice navigation.

Chapter 6 considers the route design problem of a pharmaceutical warehouse. The chapter presents a good healthcare application of logistics problems over real data obtained from a pharmaceutical logistics company.

Another application in healthcare logistics is presented in Chapter 7 in which medical suppliers are evaluated through a fuzzy linguistic representation model.

Relief logistics is one of the leading research areas in logistics, due to its importance and challenges. Chapter 8 discusses logistics planning after major disasters. The chapter mainly focuses on connecting the underlying road network of a region so that relief materials can be transported. The chapter also provides a case study of the Istanbul highway network.

Chapter 9 represents a public downtown transportation system for one of the major cities of Turkey. The chapter discusses a real data-driven simulation model that outputs a new shuttle system with fewer buses.

Another application-oriented study is discussed in Chapter 10 in which a relocation problem of a multinational electronics and electrical engineering company is solved. Utilizing real material flows and departmental relationships, it develops new layout options.

Chapter 11 focuses on continuous facility location problems and provides a comprehensive review. The chapter details a synthesis based on objective functions, distance measures, problem types, and solution methodologies.

For applications involving consolidation, logistics problems often arise in connection with scheduling and/or lot sizing decisions. Chapter 12 discusses such an example and a model that integrates routing and batching problems.

Finally, Chapter 13 considers joint replenishment and transportation problems. Especially in cargo-handling logistics operations at seaports and container terminals, transportation costs have a differentiable but nonlinear structure. Chapter 13 discusses such functions, which also have substitutable inputs, and provides an analysis utilizing dynamic and mixed integer formulations.

1

DAILY PLANNING FOR THREE-ECHELON LOGISTICS ASSOCIATED WITH INVENTORY MANAGEMENT UNDER DEMAND DEVIATION

YOSHIAKI SHIMIZU

Contents

1.1 Introduction

Due to service innovation and pressures to improve agility and greenness, daily logistics optimization is becoming important in Japan, especially for small businesses like convenience stores and supermarkets. A recent review of articles published on supply chain management within the last decade has revealed a scarcity of models that capture dynamic aspects relevant to real-world applications and has underscored the need for extensive studies on this topic (Melo et al., 2009).

In view of these observations, in this chapter, we investigate three-echelon logistic network optimization and provide a practical hybrid metaheuristic method. The model supports decision making at the tactical level for daily planning and inventory management in the presence of demand deviation. To deal with this problem, we extend our strategic approach to include some decisions at the operational level. In particular, we consider the multivehicle routing problem (M-VRP) while taking into account inventory management issues.

By taking into account the dynamics of demand and warehouse inventory, we try to give a practical approach that can provide innovative resolutions to daily planning problems. Then, to examine some effects of demand deviation on inventory condition, we carried out a parametric study regarding ordering points. The final aim of this study is to develop an integrated information and decision support system (DSS) that can dynamically manage appropriate databases of resources and product demand (see Figure 1.1). Additional work must be undertaken to realize this goal, such as the deployment of variants of the basic idea and the use of parallel computation to increase the speed of finding solutions and to enhance information retrieval and the visualization of results on a real map. We also describe our efforts along these lines in this chapter.

1.2 Problem Statements

1.2.1 Background of the Study

Noticing the growing importance of logistics network optimization, as mentioned earlier, we have investigated specific problems and the overall framework for solving such problems by considering the

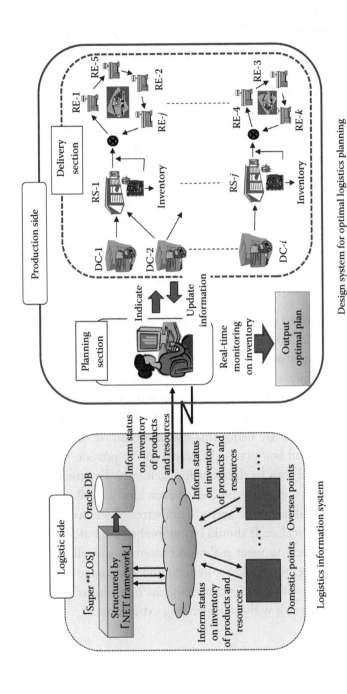

Figure 1.1   Global overview of a DSS for logistics planning.

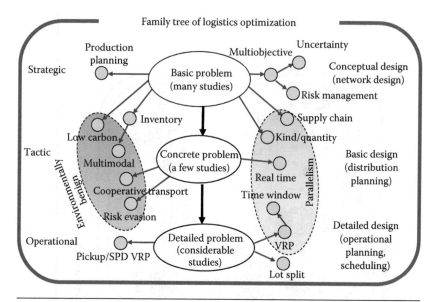

Figure 1.2   Family tree of logistics optimization problems.

similarity of problem classes in logistics systems and production planning (Shimizu, 2011a). We noticed that logistics optimization problems
are broadly classified as strategic (network design), tactical (distribution
planning), and operational (operational planning), just as production
planning problems are. We also find a similarity to design tasks for artificial products, which are classified into three levels: conceptual, basic,
and detailed designs. By using this classification, we should be able
to define the system boundary adequately and to provide the required
information, which can be different for each level. Eventually, we will
have a suite of ideas that can be used at each level as well as across the
levels. Figure 1.2 illustrates a sort of family tree of techniques that we
are attempting to realize. It should be comprehensive but involve some
cutting-edge aspects to meet real-world interests as well. It becomes
essential, therefore, to develop a general and systematic approach so
that we can cope with a variety of problems in a similar manner and
reach the desired goals without expending extra effort.

1.2.2 Brief Review of Related Studies

The strategic problem class described in the previous section has been
studied for a long time. Problems in this class are often formulated

mathematically as mixed-integer programming problems that are nondeterministic polynomial time hard (NP-hard). Hence, the number of studies is still growing by virtue of the outstanding advancements in both computer software and computer hardware. Nevertheless, hybrid methods have an advantage over these techniques, because it is almost impossible to solve real-world problems by using commercial solvers. Moreover, we note that transportation cost accounting is reasonably described by a bilinear model of loading weight and travel distance (Ton–Kilo or Weber basis).

There are also many studies belonging to the operational level. A popular problem studied at this level is named vehicle routing problem (VRP; Yeun et al., 2008). VRP is an NP-hard combinatorial optimization problem on minimizing the total distance traveled by a fleet of vehicles under various constraints. This transportation of goods from depots to all customers must be considered under the constraint that each vehicle must take a circular route with the depot as its starting point and destination.

Recent studies of VRP can be roughly classified in the following ways. One type is an extension from generic customer demand satisfaction and vehicle payload limit conditions to include practical concerns, such as customer availability or time windows (Hashimoto et al., 2006; Mester et al., 2007), pickups (Gribkovskaia et al., 2007), and split and mixed deliveries (Mota et al., 2007). These extensions are considered both separately and in combination (Zhong and Cole, 2005). The second type is known as the multidepot problem, in which deliveries can originate from multiple depots (Wu et al., 2002; Chen et al., 2005; Crevier et al., 2007). The third type investigates multiobjective formulations of the single-depot and multidepot problems (Murata and Itai, 2005; Pasia et al., 2007; Jozefowiez et al., 2008; Geiger, 2010). Recently, many researchers have been interested in VRP with varying pickup and delivery configurations because this is the most practical and suitable way to consider reverse logistics (Min, 1989; Catay, 2010; Goksal et al., 2013). These studies can be classified into three categories (Nagy and Salhi, 2005): delivery first and pickup second, in which pickup happens only after delivery; mixed pickup and delivery (VRPMPD), in which delivery and pickup are permitted in any sequence along the routes; and simultaneous pickup and delivery (VRPSPD). The VRPSPD problem reduces to the VRPMPD problem if only either pickup or delivery is required at each customer. Instances of the VRPSPD problem are frequently encountered in the distribution system of bottled drinks, groceries, liquid propane gas tanks, hotel laundry services, etc. Due to the difficulty of solving such problems, only small instances of VRP are solved to validate the effectiveness of the approaches. Moreover, it should be noted that those studies consider only distance (Kilo basis) to derive the route. To solve VRP in terms of the Ton–Kilo basis, we developed a hybrid approach composed of a modified saving method and modified tabu search (Shimizu, 2011b,c).

Apart from those two classes of problem, there have been a few studies (Tuzun and Burke, 1999; Albareda-Sambola et al., 2005; Prins et al., 2006; Zhao et al., 2008) at the tactical level. At this level, it is necessary to consider connections to both the upper (strategic) level and the lower (operational) level. In such problems, decisions about allocations to the depot are considered in addition to VRP. Hence, we must take care to use consistent transport cost accounting. However, it is common to use the Ton–Kilo basis at the strategic level and the Kilo basis at the operational level. Moreover, each formulated problem is NP-hard. Thus, it becomes necessary to resolve the inconsistency in cost accounting while coping with the inherent hardness of the problem.

1.3 Problem Formulation

For a global logistics network composed of major distribution centers (DCs), sub-DCs (i.e., depots) (RSs), and customers (REs), we wish to determine the available depots, paths from DCs to depots, and circular routes from every depot to its client customers (refer to *Delivery section* in Figure 1.1). The goal of this problem is to minimize the total cost for daily logistics over planning horizon *T*. This problem is formulated as the following mixed-integer programming problem under some mild assumptions: round-trip transport between DC and depot, unimodal transport, averaged time-invariant unit costs and system parameters (except for demand and inventory), independence (separability) of decisions per planning period, and so forth.

$(p.1)$ Minimize for every $t \in T$

$$\sum_{i \in I} \sum_{j \in J} \left(C_{ij} d1_{ij} + Hp_i \right) f_{ij}(t) + \sum_{v \in V} \sum_{p \in P} \sum_{p' \in P} c_v d2_{pp'} \left(g_{pp'v}(t) + q_v \right) z_{pp'v}(t)$$

$$+ \sum_{j \in J} \left(Ho_j r_j(t) + Ha_j s_j(t) + Hs_j \left(s_j(t) + \sum_{i \in I} f_{ij}(t) \right) \right) + \sum_{v \in V} F_v y_v(t)$$

subject to the constraints

$$\sum_{p \in P} z_{kpv}(t) \leq 1, \quad \forall k \in K; \forall v \in V, \quad \exists t \in T \tag{1.1}$$

$$\sum_{p' \in P} z_{pp'v}(t) - \sum_{p' \in P} z_{p'pv}(t) = 0, \quad \forall p \in P; \forall v \in V, \quad \exists t \in T \tag{1.2}$$

$$\sum_{j' \in J} z_{jj'v}(t) = 0, \quad \forall j \in J, \forall v \in V, \quad \exists t \in T \tag{1.3}$$

$$s_j(t) + \sum_{i \in I} f_{ij}(t) = \sum_{v \in V} \sum_{k \in K} g_{jkv}(t), \quad \forall j \in J, \quad \exists t \in T \tag{1.4}$$

$$r_j(t) + s_j(t) + \sum_{i \in I} f_{ij}(t) \leq Q_j x_j(t), \quad \forall j \in J, \quad \exists t \in T \tag{1.5}$$

$$g_{pp'v}(t) \leq W_v z_{pp'v}(t), \quad \forall p \in P; \forall p' \in P; \forall v \in V, \quad \exists t \in T \tag{1.6}$$

$$\sum_{p \in P} \sum_{p' \in P} z_{pp'v}(t) \leq M y_v(t), \quad \forall v \in V, \quad \exists t \in T \tag{1.7}$$

$$\sum_{k \in K} g_{kjv}(t) = 0, \quad \forall j \in J; \forall v \in V, \quad \exists t \in T \tag{1.8}$$

$$\sum_{v \in V} \sum_{p \in P} g_{pkv}(t) - \sum_{v \in V} \sum_{p \in P} g_{kpv}(t) = D_k(t), \quad \forall k \in K, \quad \exists t \in T \tag{1.9}$$

$$\sum_{p \in P} \left(g_{pkv}(t) - D_k(t) z_{pkv}(t) \right) = \sum_{p \in P} g_{kpv}(t), \quad \forall k \in K, \forall v \in V, \exists t \in T$$

$$\tag{1.10}$$

$$\sum_{j \in J} \sum_{k \in K} z_{jkv}(t) = y_v(t), \quad \forall v \in V, \quad \exists t \in T \qquad (1.11)$$

$$\sum_{j \in J} \sum_{k \in K} z_{kjv}(t) = y_v(t), \quad \forall v \in V, \quad \exists t \in T \qquad (1.12)$$

$$\sum_{p \in \Omega} \sum_{p' \in \Omega} z_{pp'v}(t) \le |\Omega| - 1, \; \forall \Omega \subseteq P \setminus \{1\}, \; |\Omega| \ge 2, \; \forall v \in V, \; \exists t \in T$$

$$(1.13)$$

$$P_i^{\min} \le \sum_{j \in J} f_{ij}(t) \le P_i^{\max}, \quad \forall i \in I, \quad \exists t \in T \qquad (1.14)$$

$$r_j(t) + s_j(t) \le S_j x_j(t), \quad \forall j \in J, \quad \exists t \in T \qquad (1.15)$$

$$x_j(t) \in \{0,1\}, \quad \forall j \in J, \forall t \in T; \quad y_v(t) \in \{0,1\}, \quad \forall v \in V, \forall t \in T$$

$$s_j(t) \ge 0, \quad \forall j \in J, \forall t \in T; \quad r_j(t) \ge 0, \quad \forall j \in J, \forall t \in T$$

$$z_{pp'v}(t) \in \{0,1\}, \quad \forall p \in P; \forall p' \in P; \forall v \in V, \quad \forall t \in T$$

$$f_{ij}(t) \ge 0, \quad \forall i \in I; \forall j \in J, \forall t \in T;$$

$$g_{pp'v}(t) \ge 0, \forall p \in P; \forall p' \in P; \forall v \in V, \forall t \in T$$

Variables

$f_{ij}(t)$: Load from DC i to depot j at time t

$g_{pp'v}(t)$: Load of vehicle v on the path from $p \in P$ to $p' \in P$ at time t

$r_j(t)$: Takeover inventory at depot j at time t

$s_j(t)$: Consumption quantity from inventory at depot j at time t

$x_j(t) = 1$ if depot j is open at time t; otherwise 0

$y_v(t) = 1$ if vehicle v is used at time t; otherwise 0

$z_{pp'v}(t) = 1$ if vehicle v travels on the path from $p \in P$ to $p' \in P$ at time t; otherwise 0

Parameters

C_{ij}: Transportation cost per unit load per unit distance from DC i to depot j

c_v: Transportation cost per unit load per unit distance of vehicle v

$D_k(t)$: Demand of customer k at time t

$d1_{ij}$: Path distance between $i \in I$ and $j \in J$
$d2_{pp'}$: Path distance between $p \in P$ and $p' \in P$
F_v: Fixed cost for the working vehicle v
Ha_j: Handling cost per unit load at depot j
Ho_j: Holding cost per unit load at depot j
Hp_i: Shipping cost per unit load from DC i
Hs_j: Shipping cost per unit load from depot j
M: Auxiliary constant (a large integer)
P_i^{max}: Maximum load available at DC i
P_i^{min}: Minimum load required to ship from DC i
q_v: Unladen weight of vehicle v
Q_j: Maximum capacity at depot j
S_j: Maximum inventory at depot j
W_v: Maximum capacity of vehicle v

Index set
  I: DC
  J: Depot
  K: Customer
  V: Vehicle
  $P = J \cup K$
  T: Planning horizon
  Ω: Subtour candidate

In (p.1), the objective function is composed of round-trip transportation costs between each DC and the opening depot (hereinafter just *the depot*); circular transportation costs for traveling to every customer; shipping costs at each DC; holding, handling, and shipping costs at each depot; and fixed costs for the working vehicles. Several constraints are applied: vehicles cannot visit a customer twice (Equation 1.1), a vehicle visiting a certain depot or customer must leave it (Equation 1.2), no direct travel between DCs (Equation 1.3), material balance (Equation 1.4), upper-bound capacity at depot (Equation 1.5), upper-bound load capacity for a vehicle (Equation 1.6), each vehicle must travel on a certain path (Equation 1.7), vehicles return to the depot empty (Equation 1.8), customer demand is satisfied by a certain vehicle (Equation 1.9), the sum of incoming goods must be greater than the outgoing goods due to demand (Equation 1.10), each vehicle leaves only one depot and returns there (Equations 1.11 and 1.12), subtour elimination constraint (Equation 1.13), the amount of goods available from DC is bounded (Equation 1.14), and the amount of inventory is bounded above (Equation 1.15). We also assume the following inventory control policy:

$$r_j(t) = \begin{cases} (1-\zeta)r_j(t-1) & \text{if } (1-\zeta)r_j(t-1) \geq R_j \\ S_j - (1-\zeta)r_j(t-1) & \text{if } (1-\zeta)r_j(t-1) < R_j, \quad \forall j \in J, \forall t \in T \end{cases}$$

(1.16)

Here, ζ (<1) and R_j are the fouling rate of unsold goods and the ordering point at depot j, respectively.

We know that it is almost impossible to solve this problem under realistic sizes using any currently available commercial software. Hence, we try to solve the problem in a hybrid manner that divides it into subproblems and applies a suitable method to each. Previously, we have combined tabu search (Glover, 1989) for the location subproblem and a graph algorithm for the allocation subproblem to develop a method called hybrid tabu search (HybTS) (Shimizu and Wada, 2004; Wada and Shimizu, 2006), and we have successfully used this approach to solve complicated logistics optimization problems arising from a variety of real-world situations. HybTS is a two-level solution method in which the upper-level subproblem optimizes the selection of available depots while the lower-level subproblem optimizes the paths from DCs to customers via depots in a way that minimizes the total cost. It is not only a practical and powerful method, but also flexible and suitable for dealing with a variety of extensions, as shown in Figure 1.2. Hence, we use a similar idea here to solve this problem in a way that is computationally effective (Shimizu and Fatrias, 2013).

1.4 Daily Decision Associated with Inventory Conditions

1.4.1 Multilevel Approach Incorporating Vehicle-Routing Problem

For daily logistics optimization, it is meaningful to take into account the inventory control at each depot. To make the hierarchical approach suitable for the present case, we have developed two new ideas and integrated them into the framework of our hybrid method. To the best of our knowledge, such a global approach has not been reported elsewhere.

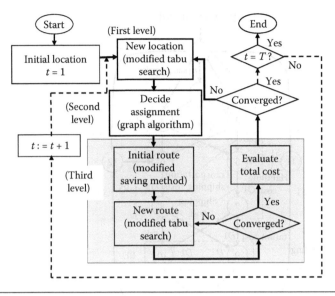

Figure 1.3   Flow chart of the solution procedure.

In the first level, we choose the available depots by a modified tabu search. Then, in the second level, we obtain tentative round-trip paths from DCs to customers via depots by a graph algorithm for solving the minimum cost flow (MCF) problem. Using the customers thus allocated as the clients for each depot, we derive vehicle routes for every depot by using the modified savings method and modified tabu search. The obtained result is fed back to the first level to evaluate another candidate set of available depots. This procedure is repeated until a given convergence condition has been satisfied. The algorithm is illustrated in Figure 1.3.

In developing this algorithm, we need to obtain the MCF graph that takes into account the inventory at each depot. For example, the case where $|I| = |J| = |K| = 2$ is illustrated in Figure 1.4. In Table 1.1, we summarize the information required for the edges and nodes in the graph. In terms of the MCF graph thus derived, we can solve the original allocation problem extremely quickly by a graph algorithm such as RELAX4 (http://mit.edu/dimitrib/www/home. html) together with its sensitivity analysis. The sensitivity analysis allows the problem to be repeatedly solved with slightly different parameters. At the end of this procedure, we can efficiently allocate

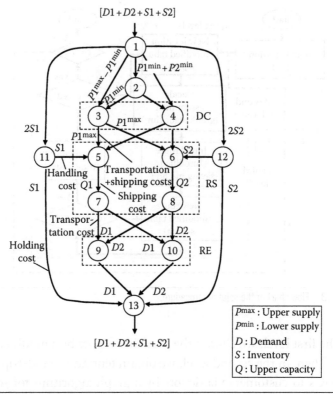

Figure 1.4 Example of an MCF graph. *Note:* Each digit refers to suffix in Table 1.1.

Table 1.1 Labeling on the Edges of an MCF Graph

EDGE (FROM–TO)	COST	CAPACITY	CASE IN FIGURE 1.4
Source–Σ(Dummy node)	$-M$	$\sum_{i \in I} P_i^{min}$	#1–#2
Source–DC i	0	$P_i^{max} - P_i^{min}$	#1–#3, #1–#4
Σ–DC i	0	P_i^{min}	#2–#3, #2–#4
DC i–RS j	$C_{ij}d1_{ij} + Hp_i$	P_i^{max}	#3–#5, #3–#6, etc.
Between double nodes of RS j	Hs_j	Q_j	#5–#7, #6–#8
Stock–RS j	Ha_j	S_j	#11–#5, #12–#6
Source–Stock j	0	$2S_j$	#1–#11, #1–#12
Stock j–Sink	Ho_j	S_j	#11–#13, #12–#13
RS j–Customer k	$c_v d2_{jk}$	D_k	#7–#9, #7–#10, etc.
Customer k–Sink	0	D_k	#9–#13, #10–#13

the client customers to each depot on the Ton–Kilo basis. In other words, the original M-VRP has now been turned into multiple ordinary VRPs.

To derive the initial solution of each VRP with consistent transport cost accounting, we apply the modified savings method whose algorithm is outlined as follows:

> Step 1: Create round-trip routes from the depot to all customers. Compute the savings value $s_{ij} = (d_{0j} - d_{0i} - d_{ij})D_j + (d_{0j} + d_{i0} - d_{ij})q_v$, where D_j, q_v, and d_{ij} denote the demand at location j, the unladen weight of the vehicle v, and the distance between locations i and j, respectively (refer to Figure 1.5).
>
> Step 2: Order these pairs in descending order of savings value.
>
> Step 3: Merge the path, following the order obtained from Step 2 as long as it is feasible and the savings value is greater than $-F_v/c_v$, where F_v denotes the fixed operational cost of vehicle v. Here, we note that the inclusion of fixed operational costs for the working vehicles in practical economic evaluations is a new idea.

However, since the modified savings method derives only an approximate solution, we apply the modified tabu search to update such a solution. The modified tabu search is a variant that probabilistically accepts even a degraded candidate in its local search, where a neighboring solution is generated from a randomly selected insert, swap, or two-opt operation. For this purpose, we applied the Maxwell–Boltzmann

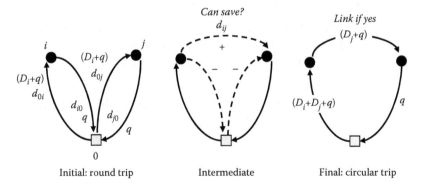

Figure 1.5 Illustrative steps to derive savings value.

probability function used in simulated annealing (Kirkpatrick et al., 1983). Here, we emphasize that the advantage of this approach is that transport cost accounting is on the same Ton–Kilo basis for the procedures at the upper levels (first and second) and the lower level (third) in Figure 1.3.

1.4.2 Analysis of Inventory Level on Demand Variation

It is commonly known that too much inventory lowers economic efficiency while the stock-out condition or opportunity loss will happen in the opposite case. For daily logistics, therefore, it is of special importance to correctly estimate the demand and properly manage the inventory. Generally speaking, although estimating demand correctly is almost impossible in many cases, it is possible to give a rough estimate of the variability from prior experience.

Under such circumstances, it is relevant and practical to try to discern the relation between demand variability and inventory level by a parametric approach. Through such analyses, we can set up a reliable inventory level to maintain economically efficient logistics while avoiding the stock-out state. Though such considerations are able to reveal many prospects for robust and reliable logistic systems, it has not been used much until now in the network optimization of logistics due to computational difficulties.

1.5 Numerical Experiments

1.5.1 Setup of Test Problem

To examine the performance of the proposed method, we considered several benchmark problems of different problem sizes (i.e., different specifications of $\{|I|, |J|, |K|\}$). Every system parameter is set randomly within a prescribed interval, as summarized in Table 1.2. The location of every member is also generated randomly, and distances between them are given by the Euclidian distance.

1.5.2 Results for the Reference Conditions

We randomly changed the demand to an amount within $(100 \pm \rho)\%$ of the demand on the previous day. Unsold goods at each depot are

Table 1.2 Notes on Parameter Setup

MEMBER	ITEM	RANGE	REMARKS
DC	Hp: Shipping cost	$100 \times [0.2, 0.8]$	<3>
	P^{max}: Available (max)	$1000 \times [0, 1] + P^{min}$	<5>, Total P^{max} > Total capacity of RS
	P^{min}: Available (min)	$1000 \times [0.2, 0.8]$	<5>, Total P^{min} > Total demand
RS	Hs: Shipping cost	$100 \times [0.2, 0.8]$	<3>
	Ha: Handling cost	$50 \times [0.2, 0.8]$	<3>
	Ho: Holding cost	$100 \times [0.2, 0.8]$	<5>
	Q: Capacity	$p \times [0.2, 0.8]$	<5>, $p = 100 \times |K|/|J|$
	S: Allowable inventory	$x \times [0.5, 0.7]$	<3>, Varying at each time
RE	D: Demand	$100 \times [0.2, 0.8]$	<3>, Total demand < Total capacity of RS[a]

Note: $C_{ij} = 3$, $c_v = 1$, $W_v = 500$, $F_v = 50,000$, and $q_v = 10$; <*n*> multiple of n.
[a] Under this condition, the stock-out status will not occur.

stored as inventory, and it is possible to use them in the following days. However, it is assumed that up to the rate (ζ) of the goods are spoiled at random and that goods are restocked to the upper limit when the inventory level falls below the prescribed safety level ($R_j = \sigma S_j$), which is equivalent to adopting a fixed-order-quantity policy.

First, we solved smaller problems, such as those characterized by $|I| = 3$, $|J| = 10$, and $|K| = 100$ over 30 days and $|I| = 5$, $|J| = 20$, and $|K| = 200$ over 10 days. Parameters ρ, σ, and ζ are set at 0.3, 0.5, and 0.1, respectively. Figures 1.6 and 1.7 illustrate the changes in demand and inventory during the planning horizon. Under these conditions, we derive the optimal cost, which broadly changes in accord with demand fluctuation, as shown in Figure 1.8. In Figure 1.9, we can see that the change in the number of active depots is moderated and kept nearly constant (around 60%). However, the activity rates of

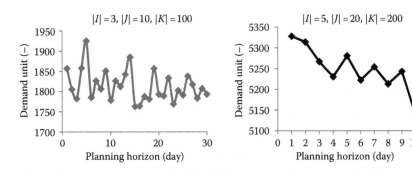

Figure 1.6 Variation of demand.

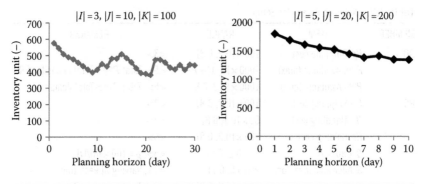

Figure 1.7 Variation of inventory.

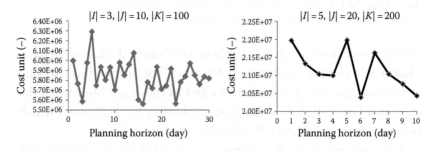

Figure 1.8 Trend of optimal cost.

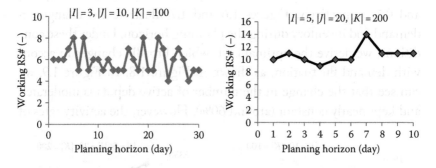

Figure 1.9 Variation of the number of active depots.

each depot differ greatly, as shown in Figure 1.10. At the next stage of logistical restructuring, the depots that have a low activity rate may be merged with those that have higher rates.

We solved some larger problems to examine the necessary computation time. Fixing the planning horizon at 1, and fixing $|I| = 10$ and $|J| = 30$, we solved the problems given by $|K| = \{250, 500,$

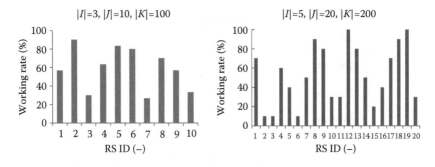

Figure 1.10   Activity rate of each RS.

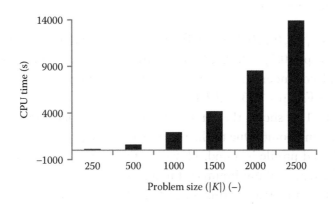

Figure 1.11   Profile of CPU time by problem size.

1000, 1500, 2000, 2500}. As expected a priori, the required CPU time increases exponentially with the size, as shown in Figure 1.11. Even for these larger problems, however, we can obtain the result within a reasonable time of around several hours.

From the convergence profile for the largest problem, shown in Figure 1.12, we can confirm that sufficient convergence is obtained. From all of these results, we can claim that the proposed method is significant and computationally effective.

1.5.3 Results over a Wide Range of Deviations

To analyze the effect of demand variation on the inventory condition, we carried out a parametric study varying the ordering points in a small model such as $|I| = 2$, $|J| = 5$, and $|K| = 100$ over 30 days. We solved every problem for each pairing of five different ordering

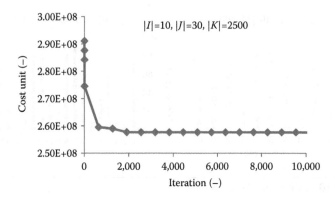

Figure 1.12   Profile of convergence.

points ($\sigma = \{0.1, 0.2, 0.3, 0.4, 0.5\}$) and four different ranges of demand variation ($\rho = \{0.2, 0.3, 0.4, 0.5\}$). In all, 600 optimization problems were solved under the same conditions as before. The results are shown in Figures 1.13 and 1.14.

Figure 1.13 shows the total cost for ranges of demand deviation and ordering point. Due to the nondeterministic parameter setting, a complicated profile is found. However, the overall shape is plausible since the region of minimum cost moves to a higher ordering point as the deviation increases. This suggests that, in terms of cost management, it is important to control the ordering point or inventory level according to the demand variation. When we separate the inventory cost from the total cost, its changes are rather simple, as shown in Figure 1.14. Since a higher stock level incurs a greater holding cost, the cost increases proportionally with the ordering point regardless of the variability of demand.

Finally, from these parametric studies, we claim that the applied model behind the mathematical formulation is adequate. The plausibility of the results supports the viability of the approach if it were used in a real-world optimization with actual parameters.

1.6 Prospects for Further Applications

1.6.1 Variants of the Modified Savings Method

We can generalize the foregoing Weber basis by introducing the power model of weight and distance. For this generalized Weber basis, the savings value for delivery is given by

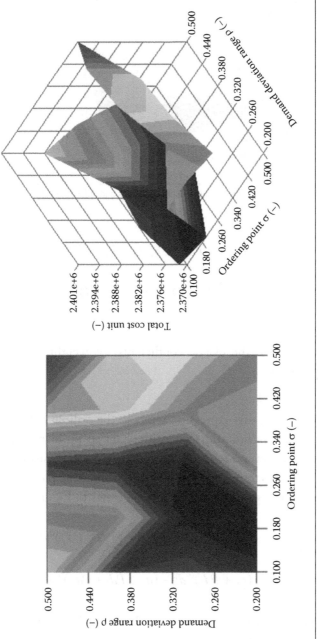

Figure 1.13   Total cost for various demand deviations and ordering points.

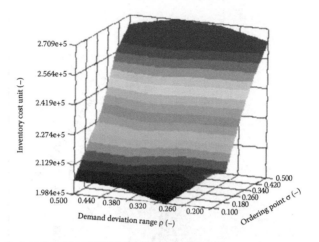

Figure 1.14 Inventory cost for various demand deviations and ordering points.

$$\frac{s_{ij}}{(\gamma c_v)} = (q_v + D_i)^\alpha d_{0i}^\beta + q_v^\alpha d_{i0}^\beta + (q_v + D_j)^\alpha d_{0j}^\beta$$

$$- (q_v + D_i + D_j)^\alpha d_{0i}^\beta - (q_v + D_j)^\alpha d_{ij}^\beta \qquad (1.17)$$

where

α and β denote the elastic coefficients for weight and distance, respectively

γ is a constant

When $\alpha = \beta = \gamma = 1$, this expression refers to the ordinary Weber basis.

Moreover, it is common to consider pickup problems instead of delivery ones in reverse logistics. Here, every vehicle visits the pickup points and returns to the depot directly. Letting Pd be pickup demand, we can derive the savings value as follows (refer to Figure 1.15):

$$\frac{s_{ij}}{(\gamma c_v)} = q_v^\alpha d_{0j}^\beta + (q_v + Pd_i)^\alpha d_{i0}^\beta + (q_v + Pd_j)^\alpha d_{j0}^\beta$$

$$- (q_v + Pd_i)^\alpha d_{ij}^\beta - (q_v + Pd_i + Pd_j)^\alpha d_{j0}^\beta \qquad (1.18)$$

The previous idea can be extended to the case where vehicles stop at an intermediate destination before returning to the depot. This is the case, for example, when a vehicle visits a disposal site to dump waste

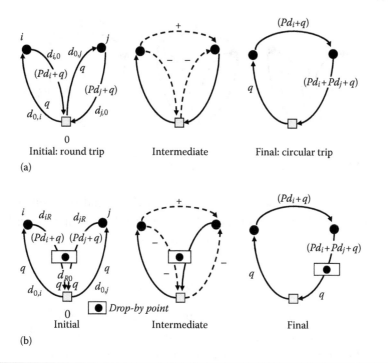

Figure 1.15 Scheme to derive savings value for pickup VRP for either (a) direct or (b) drop-by pickup.

or a remanufacturing facility to deliver used products. In this case, we can modify this equation as follows:

$$\frac{s_{ij}}{(\gamma c_v)} = (q_v + Pd_i)^\alpha d_{iR}^\beta + q_v^\alpha d_{R0}^\beta + q_v^\alpha d_{0j}^\beta + (q_v + Pd_j)^\alpha d_{jR}^\beta$$

$$- (q_v + Pd_i)^\alpha d_{ij}^\beta - (q_v + Pd_i + Pd_j)^\alpha d_{jR}^\beta \qquad (1.19)$$

where the subscript R denotes the intermediate destination.

Similarly, we have the following expression in the case of VRPSPD (refer to Figure 1.16):

$$\frac{s_{ij}}{(\gamma c_v)} = (D_i + q_v)^\alpha d_{0i}^\beta + (Pd_i + q_v)^\alpha d_{i0}^\beta + (D_j + q_v)^\alpha d_{0j}^\beta$$

$$+ (Pd_j + q_v)^\alpha d_{j0}^\beta - (D_i + D_j + q_v)^\alpha d_{0i}^\beta$$

$$- (Pd_i + D_j + q_v)^\alpha d_{ij}^\beta - (Pd_i + Pd_j + q_v)^\alpha d_{j0}^\beta \qquad (1.20)$$

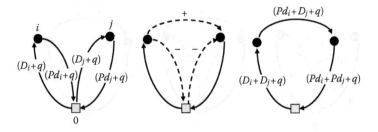

Figure 1.16   Scheme to derive savings value for VRPSPD.

Just by applying these formulas to derive the initial solution of the respective VRP, we can simply follow the overall procedure described earlier to obtain the respective final solutions.

1.6.2 Application of Parallel Computing Techniques

Unlike conventional studies, our method can cope with large-scale problems in a practical and flexible manner. However, there still exists a great need for solving problems more quickly and efficiently in order to make responsive decisions in global markets. For such requirements, it is natural to develop a parallel implementation for logistics optimization. In a special issue of *Parallel Computing*, Laporte and Musmanno (2003) emphasized the importance of parallel computing in logistics not only due to the large scale of these problems but also because of real-time applications arising in the delivery of emergency services and in courier or dial-a-ride services.

To make our hybrid approach suitable for parallel optimization, we developed a binary particle swarm optimizer (PSO) and substituted it for the modified tabu search used in the previous procedure. Compared with an individual search such as tabu search, the population-based PSO algorithm is well suited to implementing a parallel computation (Shimizu and Ikeda, 2010). The first application considered a rather simplified formulation for strategic problems using master–worker parallelism, as illustrated in Figure 1.17. Then, more complicated configurations were examined. For these, a novel parallel procedure similar to the island model used in genetic algorithms was developed, employing multithreading techniques so that the idle time for the parallel computation becomes very small. Moreover, the effect of the topology of subpopulations (Figure 1.18) and the manner

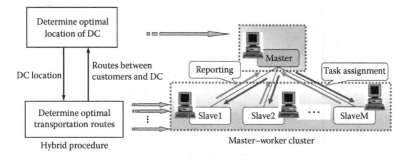

Figure 1.17   Scheme under master–worker parallelism.

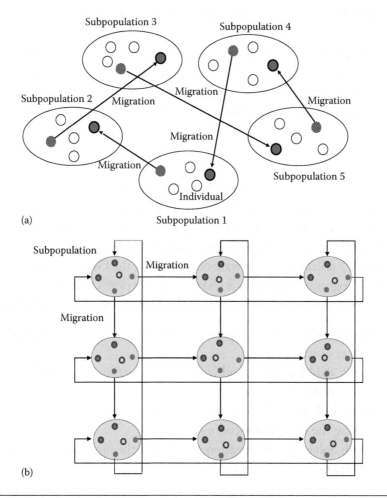

Figure 1.18   Topology for island model parallelism: (a) random ring (RR) or (b) two-node torus ring (2nR).

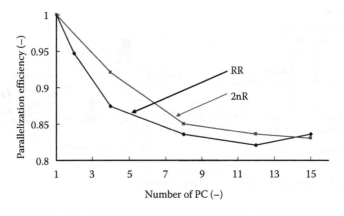

Figure 1.19 Efficiency of parallelization against the number of processors.

of information exchange between subpopulations were analyzed. Finally, we showed that such an approach can solve huge problems with more than 20,000 customers while maintaining high efficiency of the parallelization, as shown in Figure 1.19.

1.6.3 Enhancement for Practical Use

To realize the planning system illustrated in Figure 1.1, it is essential to provide a user-friendly interface to manage the system. In the planning section on the production side, this goal is closely related to data handling and visualization of the circumstances at hand. For this, we can effectively utilize some software developed for the Google Maps application programming interface (API). We have developed the following stepwise procedure by using JavaScript and appropriate free software:

Step 1: Collect the addresses of locations in an Excel spread sheet or text file.
Step 2: Add longitude and latitude information for every location in the sheet.
Step 3: Calculate the distance between every pair of locations by using the Google geocoding API.
Step 4: Solve the optimization problem by the proposed method.
Step 5: Display the routes obtained from Step 4 in Google Maps.

Figure 1.20 shows some results for an illustrative problem, in which every depot has a single route, $|I| = 1$, $|J| = 3$, and $|K| = 17$. In this figure,

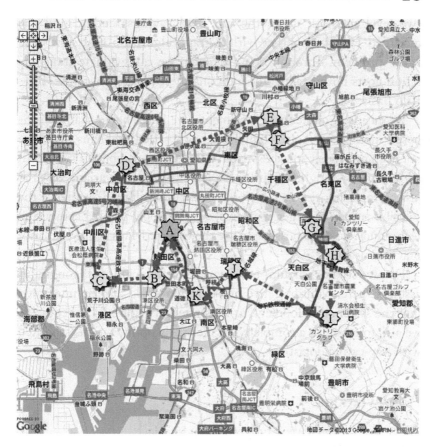

Figure 1.20   A part of the display of a result (Route from depot 1 [marked as A]).

for simplicity, the routing paths from only depot 1 are shown. Marks for locations (A–B–·····–K–A) and dotted arrows are superimposed to help visualize the actual circular route. We can see that this kind of visual information is very helpful for some tasks at an operational level. However, there still remain many possibilities to add more and valuable service information from geographical information system (GIS) applications and the Google Maps API.

1.7 Conclusion

We have described a hierarchical approach to optimize daily logistics including inventory control at depots and vehicle routing for customer delivery. For this purpose, we have extended our existing two-level method, using the modified savings method and the modified HybTS together with a graph algorithm that solves the MCF problem. Through this approach, we can evaluate transportation costs both practically and consistently in terms of the Ton–Kilo basis.

By means of numerical experiments, we have shown that the proposed method can solve complicated and varied problems that have not been previously solvable within a reasonable computation time. In addition, it is straightforward to apply the method to variants of VRP just by replacing the savings value in the procedure. To enhance the solution speed for larger problems, we can apply parallel computing techniques. It is also possible to use the Google Maps API to enhance practical usability.

Future studies should be devoted to relaxing the conditions assumed here. Multiobjective optimization could also be integrated into the system development, as illustrated in Figure 1.1. Eventually, we aim to establish a complete DSS for daily optimization associated with low-carbon logistics.

Abbreviations

API	Application programming interface
CPU	Central processing unit
DC	Distribution center
DSS	Decision support system
GIS	Geographical information system
HybTS	Hybrid tabu search
MCF	Minimum cost flow
M-VRP	Multivehicle routing problem
NP-hard	Nondeterministic polynomial time hard
PSO	Particle swarm optimization
RS	Relay station of DC, or depot
RE	Retailer, or customer
RELAX4	Software name for MCF problem
VRP	Vehicle routing problem
VRPSPD	VRP with simultaneous pickup and delivery
VRPMPD	VRP with mixed pickup and delivery

References

Albareda-Sambola, M., J. A. Diaz, and E. Fernandez. 2005. A compact model and tight bounds for a combined location-routing problem. *Computers & Operations Research* 32:407–428.

Catay, B. 2010. A new saving-based ant algorithm for the vehicle routing problem with simultaneous pickup and delivery. *Expert Systems with Applications* 37:6809–6817.

Chen, S., A. Imai, and B. Zhao. 2005. A SA-based heuristic for the multi-depot vehicle routing problem. *Japan Institute of Navigation* 113:209–216 (in Japanese).

Crevier, B., J.-F. Cordeau, and G. Laporte. 2007. The multi-depot vehicle routing problem with inter-depot routes. *European Journal of Operational Research* 176:756–773.

Geiger, M. J. 2010. Fast approximation heuristics for multi-objective vehicle routing problems. *EvoApplications*, Part II, LNCS Vol. 6025, pp. 441–450. Springer-Verlag, Berlin, Germany.

Glover, F. 1989. Tabu search—Part I. *ORSA Journal on Computing* 1:190–206.

Goksal, F. P., I. Karaoglan, and F. Altiparmak. 2013. A hybrid discrete particle swarm optimization for vehicle routing problem with simultaneous pickup and delivery. *Computers & Industrial Engineering* 65:39–53.

Gribkovskaia, I., O. Halskau, G. Laporte, and M. Vlcek. 2007. General solutions to the single vehicle routing problem with pickups and deliveries. *European Journal of Operational Research* 180:568–584.

Hashimoto, H., T. Ibaraki, S. Imahori, and M. Yagiura. 2006. The vehicle routing problem with flexible time windows and traveling times. *Discrete Applied Mathematics* 154:2271–2290.

Jozefowiez, N., F. Semet, and E.-G. Talbi. 2008. Multiobjective vehicle routing problems. *European Journal of Operational Research* 189:293–309.

Kirkpatrick, S., C. D. Gelatt, and M. P. Vecch. 1983. Optimization by simulated annealing. *Science* 220:671–680.

Laporte, G. and R. Musmanno. 2003. Parallel computing in logistics. *Parallel Computing* 29:553–554.

Massachusetts Institute of Technology, Laboratory for information and decision systems. Available from http://mit.edu/dimitrib/www/home.html (accessed May 31, 2013).

Melo, M. T., S. Nickel, and F. Saldanha-da-Gama. 2009. Facility location and supply chain management—A review. *European Journal of Operational Research* 196:401–412.

Mester, D., O. Braysy, and W. Dullaert. 2007. A multi-parametric evolution strategies algorithm for vehicle routing problems. *Expert Systems with Applications* 32:508–517.

Min, H. 1989. The multiple vehicle routing problem with simultaneous delivery and pick-up points. *Transportation Research, Part A: General* 23:377–386.

Mota, E., V. Campos, and A. Corberan. 2007. A new metaheuristic for the vehicle routing problem with split demands. *EvoCOP 2007*, LNCS Vol. 4446, pp. 121–129. Springer-Verlag, Berlin, Germany.

Murata, T. and R. Itai. 2005. Multi-objective vehicle routing problems using two-fold EMO algorithms to enhance solution similarity on non-dominated solutions. *Evolutionary Multi-Criterion Optimization*, LNCS Vol. 3410, pp. 885–896. Springer-Verlag, Berlin, Germany.

Nagy, G. and S. Salhi. 2005. Heuristic algorithms for single and multiple depot vehicle routing problems with pickups and deliveries. *European Journal of Operational Research* 162:126–141.

Pasia, J. M., K. F. Doerner, R. F. Hartl, and M. Reimann. 2007. A population-based local search for solving a bi-objective vehicle routing problem. *EvoCOP 2007*, LNCS Vol. 4446, pp. 166–175. Springer-Verlag, Berlin, Germany.

Prins, C., C. Prodhon, and R. W. Calvo. 2006. A memetic algorithm with population management (MA-PM) for the capacitated location-routing problem. *EvoCOP 2006*, LNCS Vol. 3906, pp. 183–194. Springer-Verlag, Berlin, Germany.

Shimizu, Y. 2011a. Practical optimization of large-scale logistics through hybrid meta-heuristic method. *Journal of the Society of Instrument and Control Engineers* 50:476–482 (in Japanese).

Shimizu, Y. 2011b. A meta-heuristic approach for variants of VRP in terms of generalized saving method. *Transactions of the Institute of Systems, Control and Information Engineers* 24:287–295 (in Japanese).

Shimizu, Y. 2011c. Advanced saving method to evaluate economic concern. *Transactions of the Institute of Systems, Control and Information Engineers* 24:39–41 (in Japanese).

Shimizu, Y. and D. Fatrias. 2013. Daily planning for three echelon logistics considering inventory conditions. *Journal of Advanced Mechanical Design, Systems, and Manufacturing* 7:485–497.

Shimizu, Y. and M. Ikeda. 2010. A parallel hybrid binary PSO for capacitated logistics network optimization. *Journal of Advanced Mechanical Design, Systems, and Manufacturing* 4:616–626.

Shimizu, Y. and T. Wada. 2004. Hybrid tabu search approach for hierarchical logistics optimization. *Transactions of the Institute of Systems, Control and Information Engineers* 17:241–248 (in Japanese).

Tuzun, D. and L. I. Burke. 1999. A two-phase tabu search approach to the location routing problem. *European Journal of Operational Research* 116:87–99.

Wada, T. and Y. Shimizu. 2006. A hybrid metaheuristic approach for optimal design of total supply chain network. *Transactions of the Institute of Systems, Control and Information Engineers* 19:69–77 (in Japanese).

Wu, T.-H., C. Low, and J.-W. Bai. 2002. Heuristic solutions to multi-depot location-routing problems. *Computers & Operations Research* 29:1393–1415.

Yeun, L. C., W. R. Ismail, K. Omar, and M. Zirour. 2008. Vehicle routing problem: Models and solution. *Journal of Quality Measurement and Analysis* 4:205–218.

Zhao, Q.-H., S. Chen, and C.-X. Zang. 2008. Model and algorithm for inventory/routing decision in a three-echelon logistics system. *European Journal of Operational Research* 191:623–635.

Zhong, Y. and M. H. Cole. 2005. A vehicle routing problem with backhauls and time windows: A guided local search solution. *Transportation Research, Part E* 41:131–144.

2

NEW LOCAL SEARCH ALGORITHM FOR VEHICLE ROUTING PROBLEM WITH SIMULTANEOUS PICKUP AND DELIVERY

MUSTAFA AVCI AND ŞEYDA TOPALOĞLU

Contents

2.1 Introduction

Over the last century, continuing increase in energy consumption and emissions of greenhouse gases, which mainly stems from industrial activities, has brought about serious environmental problems such as global warming and climate change. Consequently, global awareness on environmental issues has increased rapidly in recent years, making the reverse logistics activities more important. The activities in the context of reverse logistics cover recycling and reusing operations that require bidirectional flow of goods. As a result, the design of the transportation systems requiring both pickup and delivery services becomes crucial for companies to minimize transportation costs.

If the distribution and collection activities are operated separately, the transportation costs of the companies increase substantially. Besides, the operations aiming to reduce energy consumption may cause extra damage to the environment and yield no significant benefit. Therefore, all these conditions necessitate the integration of pickup and delivery operations. The vehicle routing problem with simultaneous pickup and delivery (VRPSPD) covers many reverse logistics systems containing bidirectional flow of goods. The problem can be encountered in a distribution/collection system involving a set of customers requiring delivery and pickup services simultaneously.

The VRPSPD can be formally defined as follows: let $G = (V, A)$ be a graph, where $V = \{v_0, v_1, \ldots, v_n\}$ is the set of vertices in which v_0 represents the central depot at which homogeneous vehicles are located and the other vertices represent the clients. $A = \{(v_i, v_j): v_i, v_j \in V, i \neq j\}$ is the arc set, and each arc $\{i, j\} \in A$ has a nonnegative cost c_{ij}. Each client $\{n_1, n_2, \ldots, n_n\}$ has a nonnegative demand quantity d_i and a nonnegative pickup quantity p_i. The objective of the VRPSPD is to determine a set of routes minimizing the cost and satisfying the following constraints: (1) every route starts and finishes at the central depot, (2) each client must be visited exactly once by exactly one vehicle, and (3) the total amount of goods to be carried by a vehicle cannot exceed the vehicle capacity. Mathematical models developed for the VRPSPD can be found in Dethloff (2001), Montané and Galvão (2006), and Nagy and Salhi (2005).

The VRPSPD can arise in many practical applications of reverse logistics, and constructing an effective solution strategy to the problem is one of the most critical issues for designing the transportation system effectively. In the bottled drinks industry, for example, full bottles are delivered, and empty ones that are used for recycling are collected simultaneously from the customers. Another practical application of the VRPSPD occurs on the collection of used materials such as car parts, industrial equipments, and computers for remanufacturing or dissembling operations (Zachariadis et al. 2009). Moreover, the problem can be seen at grocery stores where pallets or boxes can be collected and reused for transportation (Dethloff 2001).

From the theoretical point of view, the VRPSPD is known to be NP-hard because it is a variant of the classical VRP, which is a well-known NP-hard problem. Large-scale real-life problem instances cannot be solved efficiently by exact solution methods. Thus, heuristic and metaheuristic solution approaches have been generally implemented as a solution method for the VRPSPD since these solution methods are capable of generating high-quality solutions to this type of problems within a reasonable computation time.

In this study, we develop a local search algorithm for solving the VRPSPD. The applied methodology is constructed by hybridizing a simulated annealing (SA)–inspired algorithm with variable neighborhood descent (VND) algorithm. The SA-inspired algorithm enables the search process to explore different search regions in the search space, while VND is implemented to generate high-quality solutions from the examined regions. One of the most important features of the developed algorithm is that it is free from parameter tuning, which makes the implementation of the algorithm easier. The proposed methodology is tested on well-known VRPSPD benchmark instances derived from the literature. The computational results show that the proposed algorithm generates competitive results with the most sophisticated algorithms developed so far, for the VRPSPD.

The remainder of this chapter is organized as follows: in Section 2.2, the studies related to the VRPSPD in the literature are reviewed. In Section 2.3, we give detailed information about our solution methodology. In Section 2.4, the computational results of the proposed algorithm obtained from the benchmark instances are presented. Finally, concluding remarks and future research are presented in Section 2.5.

2.2 Literature Review

Over the last decade, VRPSPD has become an interesting topic for researchers because of its applicability on reverse logistic systems containing distribution and collection of goods. In this section, we review the exact and heuristic solution approaches proposed for the VRPSPD.

The VRPSPD is first introduced by Min (1989). In this study, a book distribution and collection problem from a central library to 22 remote libraries is handled. In order to solve the problem, customers are clustered first and then the traveling salesman problem is solved for each cluster. Dethloff (2001) developed an insertion-based heuristic algorithm consisting of four different insertion criteria: traveling distance, residual capacity, radial surcharge, and combination. Furthermore, the author presented a mathematical model for the problem and discussed the relationship between the VRPSPD and other VRP variants. Nagy and Salhi (2005) proposed an integrated heuristic approach for the VRPSPD. The algorithm comprises of various routines used for feasibility and improvement. The first exact algorithm for the VRPSPD has been developed by Dell'Amico et al. (2006). In this study, two different strategies, exact dynamic programming and state space relaxation, are used for sub pricing problem. The algorithm can solve instances up to 40 customers optimally. Gajpal and Abad (2010) presented saving-based heuristics for the VRPSPD. In these heuristics, two existing routes are merged in order to create a new route. The feasibility of the new route is checked by employing a cumulative net-pickup approach. Another exact solution approach has been suggested by Subramanian et al. (2011). The authors developed a branch-and-cut algorithm for the VRPSPD. The algorithm finds improved lower bounds and several new optimal solutions.

Metaheuristic algorithms have been widely implemented for the VRPSPD. Especially, single-solution-based algorithms based on tabu search (TS) have been commonly used. Crispim and Brandão (2005) are the first researchers to use a metaheuristic algorithm for the VRPSPD. In their study, an algorithm constructed by hybridizing TS and VND has been developed. Chen and Wu (2005) presented another hybrid metaheuristic algorithm consisting of TS and record-to-record travel algorithm. In this algorithm, initial solution is constructed by using an insertion-based algorithm, and subsequently 2-exchange, swap, shift, 2-opt, and or-opt neighborhood generation mechanisms are used to improve the initial solution. Montané and Galvão (2006) developed a TS-based heuristic approach in which shift, cross, and 2-opt routines are employed.

Bianchessi and Righini (2007) proposed constructive and local search heuristics and a TS algorithm that contains variable neighborhood structures. Wassan et al. (2008) developed a reactive TS algorithm for the VRPSPD. The developed algorithm consists of two phases: generating an initial solution and improving on it. A modified sweep algorithm is used to obtain an initial solution and the neighborhood structures that are shift, swap, local shift, and reverse, respectively, used for improvement. The authors also constructed a mechanism that dynamically controls the tabu list size to provide an effective balance between the intensification and diversification of the search. Zachariadis et al. (2009) presented a hybrid metaheuristic algorithm that combines TS and guided local search (GLS) heuristics. In this study, an initial solution is generated by a saving-based constructive heuristic, and then the solution is improved by the hybrid TS–GLS methodology with the neighborhood structures that are customer relocation, customer exchange, route interchange I, and route interchange II, respectively.

Zachariadis et al. (2010) suggested a heuristic algorithm based on adaptive memory methodology. The heuristic algorithm collects properties of good solutions found during the search process. New solutions are generated by combining these properties, and subsequently an improvement procedure based on TS is applied. Moreover, an additional memory mechanism has been proposed to provide appropriate diversification in the search. Zachariadis and Kiranoudis (2011) presented a local search algorithm that is capable of exploring rich solution neighborhoods effectively. In order to examine these neighborhood types, the authors use an algorithmic concept, called Static Move Descriptor (SMD), which statically encodes tentative moves. To diversify the search efficiently, another algorithmic framework, called promises concept, which is a variation of the aspiration criteria of TS, has been used. Ropke and Pisinger (2006) proposed a large neighborhood search (LNS) algorithm for some VRP variants involving the VRPSPD. A parallel algorithm is developed by Subramanian et al. (2010). The algorithm is embedded with a multistart heuristic that comprises VND procedure integrated in an iterated local search (ILS) framework. The developed algorithm automatically calibrates some parameters, which makes it self-adaptive, avoiding the need of manual tuning. Furthermore, the algorithm has the ability of exploring the high level of parallelism inherent to recent multicore clusters.

Recently, population-based algorithms have been also applied for the VRPSPD. Ai and Kachitvichyanukul (2009) proposed a particle swarm optimization (PSO) algorithm for solving the VRPSPD. In this study, a random key-based encoding and decoding method is applied, and 2-opt and a heuristic approach based on the cheapest insertion are implemented for improvement. Gajpal and Abad (2009) developed an ant colony optimization (ACO) algorithm for the problem. The authors used the nearest-neighbor heuristic to construct an initial solution by means of which the trail intensities and parameters are initialized. Then, the trail intensities are used to generate an ant solution for each ant. After that, a local search consisting of 2-opt, customer insertion/interchange, and sub-path exchange is implemented on each ant solution, and elitist ants and trail intensities are updated.

Another heuristic approach based on ACO is suggested by Çatay (2010). In this study, a new saving-based visibility function and pheromone updating rule are proposed. The nearest-neighbor heuristic is implemented to generate an initial solution, and a local search consisting of four routines, namely, intra-move, intra-swap, inter-move, and inter-swap, is performed to improve the solutions. Tasan and Gen (2012) presented a metaheuristic approach based on genetic algorithm. In this algorithm, a permutation-based representation is used. Initial population is constructed randomly, and genetic operators, crossover, and mutation are implemented on the members of the population. As a selection rule, the roulette wheel selection is applied. Goksal et al. (2013) proposed a hybrid metaheuristic algorithm based on PSO in which VND algorithm is implemented for local search. In the hybrid algorithm, PSO is performed to explore good solutions in the solution space, and VND is used to improve random solutions selected from the population in each iteration of the algorithm. In addition, an annealing-like strategy is applied so as to preserve the swarm diversity of the PSO. To represent a solution, the authors use the giant tour representation; by this way, the splitting procedure proposed by Prins (2004) is adapted for obtaining a feasible solution from the giant tour for the VRPSPD.

The related literature of the VRPSPD involves powerful heuristic methods, which are successfully applied to the problem. However, many of them suffer from parameter tuning. Thus, this study proposes a simple adaptive local search algorithm that does not need parameter setting because it is developed by hybridizing two parameter-free algorithms, which are an SA-inspired algorithm, self-adaptive local search (SALS), and VND.

2.3 Proposed Solution Methodology

In this section, our proposed solution methodology for the VRPSPD is introduced. Firstly, we give information about the SALS and the VND algorithms that constitute our hybridized solution method. After that, we describe the details of the developed solution methodology.

2.3.1 Self-Adaptive Local Search

SALS is an SA-inspired metaheuristic algorithm proposed by Alabas-Uslu and Dengiz (2011). The algorithm involves a nonmonotone threshold accepting function with only one generic parameter called acceptance parameter. The parameter is adjusted automatically during the search process according to the information received from the problem and performance measure of the algorithm. The main feature of the algorithm is that it never requires a parameter tuning, which simplifies its application to the optimization problems.

As we mentioned earlier, SA is a stochastic search method that has a threshold function used to escape from local optima. For a minimization problem, the search begins from an initial solution, x; it then generates a new candidate solution, x', with a specified neighbor generation method. The difference between the objective function values, $\Delta = f(x') - f(x)$, is calculated. If Δ is nonpositive, then the candidate solution is accepted as a new current solution. Otherwise, it is accepted with a probability value, $\exp^{-\Delta/T}$, where T is a control parameter that corresponds to the temperature of the annealing schedule. T generally starts from a high value and monotonically decreases throughout the search. The algorithm is terminated when T reaches a predetermined value, called final temperature.

In a similar fashion with SA, starting from an initial solution, x, SALS generates a new solution, x′, with a neighbor generation procedure at each iteration. Whether the generated solution is accepted or not is determined according to the following acceptance condition: if $f(x') \leq t \cdot f(x)$, then $x \leftarrow x'$. In this situation, t represents the self-adaptive parameter of SALS. When the candidate solution is accepted, the algorithm proceeds to the next iteration. Otherwise, a new candidate solution is generated. The value of t is increased by the equation of $t = t + a_1 \cdot a_2$ when the number of consecutive rejected solutions is equal to the neighborhood size of the current solution (Alabas-Uslu and Dengiz 2011). The only parameter of SALS, t, is updated during the search according to two criteria: the number of improved solutions obtained during the search process and the ratio of the best solution to current solution at each iteration. Therefore, two performance indicators that are obtained from the following equations are used to manipulate t:

$$a_1 = \frac{f(x_b)}{f(x)} \qquad (2.1)$$

$$a_2 = \frac{C_i}{i} \qquad (2.2)$$

In these equations, C_i represents the number of improved solutions found through the search, and x_b and x represent the best solution and the current solution at iteration i, respectively. Parameter t is calculated throughout the search process by Equation 2.3. Whenever a new improved solution is found, a_1 decreases while a_2 increases. The value of C_i gives information about the search space structure. High values of C_i indicate that the search space has many local optimum points. On the other hand, when the search space is smooth, C_i takes low values:

$$t = 1 + a_1 \cdot a_2 \qquad (2.3)$$

2.3.2 Variable Neighborhood Descent

VND is a deterministic version of the variable neighborhood search, which is a well-known local search algorithm proposed by Mladenović and Hansen (1997). The algorithm is commonly integrated with other metaheuristic algorithms in order to increase the intensification of the search. The main idea of the algorithm is to explore the search region by using different neighborhood structures successively. At the beginning of the algorithm, VND executes the local search with the first neighborhood structure $k = 1$ until no improved solutions can be found. Then, the local search continues with the second neighborhood structure $k = 2$. If an improvement has been found with the second neighborhood structure, VND returns to the first neighborhood structure to restart the search. Otherwise, it proceeds to the third structure, and so on. If the last neighborhood structure $k = k_{max}$ yields no improved solutions, VND is finished and the obtained solution is considered as a local optimum with respect to all neighborhood structures.

2.3.3 Hybridization of SALS with VND (hybrid_SALS)

As mentioned earlier, SALS is a parameter-free algorithm, and this special feature makes its integration with other heuristic methods easier. The use of a threshold function that starts from a high value diversifies the search process and allows the algorithm to explore different regions in the search space. Besides, the nonmonotone nature of the threshold function makes it easier to escape from local minima at the end of the search. Although SALS provides sufficient diversification to the search, it may require more intensification as the algorithm encounters different search regions. Thus, hybridizing SALS with another local search heuristic that leads to an intensification effect enables the solution method to obtain more qualified solutions from the search regions examined. Therefore, this chapter proposes a solution strategy based on SALS and VND. SALS is integrated with VND in a way that whenever a new improved solution is found by SALS, VND is implemented on the solution to obtain a more qualified solution. In the developed algorithm, we apply SALS for discovering different search regions in the search space, whereas VND is

```
Procedure hybrid_SALS
Input: x, iter_sals, neighborhood structures
Output: Global best solution, x_gb
N-Neighborhood structure number;
C_1←1;
i←1;
x_gb←x;
x_1b←x;
age←0;
m_1←0;
while m1<iter_sals
    a_1←f(x_1b)/f̄(x);
    a_2←C_1/i;
    t←1+a_1.a_2;
    Generate random neighbor solutions of x and select the best one as
candidate solution, x';
    if f(x')<=t.f(x_1b)
        x←x';
        i←i+1;
        age←0;
        if f(x)<f(x_1b)
            x_1b←x;
            C_1←C_1+1;
            x''←p_VND(x_1b, NS_inter);
            if f(x'')<f(x_gb)
                x_gb←x'';
            end
        else
            m_1←m_1+1;
        end
    else
        age←age+1;
        if age==N
            age←0;
            t←t+a_1.a_2;
        end
    end
end
end
```

Figure 2.1   Main steps of hybrid_SALS.

used to increase the intensification of the search process. Thus, SALS is employed as our global search technique, and VND is used as our regional search method. The main steps of the first stage of the algorithm are given in Figure 2.1, where x_{1b} represents the best solution obtained from SALS and x_{gb} represents the global best solution found after applying VND. The neighbor solutions are obtained by applying each neighborhood structure, which will be described in Section 2.3.5. Whenever x_{1b} is updated, our VND algorithm, which will be described in Section 2.3.6, is triggered for improvement. The algorithm is terminated when the total number of consecutive iterations without improvement, m_1, reaches a predetermined value that is represented as iter_sals in Figure 2.1.

Two slight modifications are applied to the original version of SALS algorithm. The first modification is implemented on the threshold function. Instead of using the objective function value of the current solution, x_{lb} is employed. Preliminary tests show that this modification produces better results than the original version. The second modification is to decrease the total number of rejected candidate solutions before increasing the value of t. In order to make the search process more sensible to the number of consecutive rejected neighbors, we increase the value of t when the total number of rejected solutions is equal to the number of neighborhood structures instead of the neighborhood size of the current solution. This modification increases the diversification of the search.

2.3.4 Solution Representation

A solution representation is given for a problem with nine customers and two vehicles in Figure 2.2. A solution x for the VRPSPD is represented as a vector that consists of a sequence of nodes, with dimension $D = n + m + 1$, where n is the number of customers and m is the number of vehicles. Node 0 represents the depot, and the

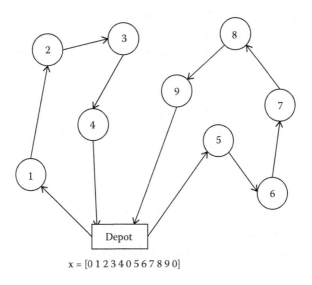

$$x = [0\ 1\ 2\ 3\ 4\ 0\ 5\ 6\ 7\ 8\ 9\ 0]$$

Figure 2.2   Solution representation.

other nodes represent customers. The solution x is represented as x = [0 1 2 3 4 0 5 6 7 8 9 0], where the first route starts from the depot and visits customers 1, 2, 3, and 4, respectively, and returns to the depot. Similarly, the second route, which is initiated and terminated at the depot, services customers 5, 6, 7, 8, and 9, respectively.

2.3.5 Neighborhood Structures

Ten neighborhood structures that are widely used for the VRP are applied in SALS and VND. Five of them are inter-route movements while others are intra-route ones. Figures 2.3 and 2.4 illustrate the intra-route and inter-route neighborhood structures, respectively.

2.3.5.1 Adjacent Swap The first intra-route neighborhood structure, Adjacent Swap, exchanges the positions of two adjacent nodes. In Figure 2.3a, the positions of adjacent nodes 2 and 3 are exchanged.

2.3.5.2 General Swap The second structure, General Swap, exchanges the positions of any node pair located in the same route. In Figure 2.3b, the positions of nodes 1 and 3 are exchanged.

2.3.5.3 Single Insertion In this structure, a node is removed and inserted between two adjacent nodes located in the same route. In Figure 2.3c, node 1 is removed and inserted between nodes 4 and 5.

2.3.5.4 Block Insertion The fourth intra-route structure, Block Insertion, removes two adjacent nodes and inserts them between another adjacent node pair. In Figure 2.3d, the adjacent nodes 1 and 2 are inserted between nodes 4 and 5.

2.3.5.5 2-Opt The final intra-route structure, 2-opt, replaces a nonadjacent arc pair with a new one, which reverses the location of nodes lying between these new arcs. In Figure 2.3e, the nonadjacent arcs (2,3) and (5,0) are deleted while the arcs (2,5) and (3,0) are inserted.

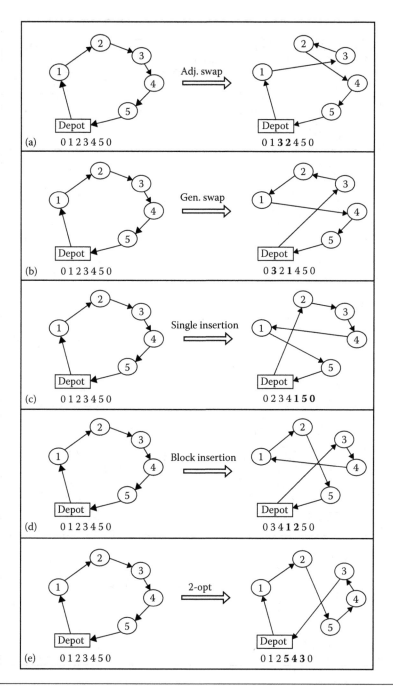

Figure 2.3 Intra-route neighborhood structures: (a) adjacent swap, (b) general swap, (c) single insertion, (d) block insertion, and (e) 2-opt.

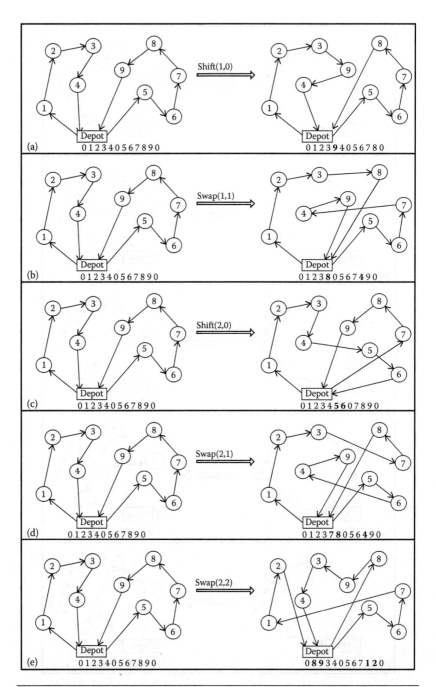

Figure 2.4   Inter-route neighborhood structures: (a) Shift(1,0), (b) Swap(1,1), (c) Shift(2,0), (d) Swap(2,1), and (e) Swap(2,2).

2.3.5.6 Shift(1,0)   The first inter-route structure, Shift(1,0), removes a customer from its current route and inserts it at a location in another route. In Figure 2.4a, node 9 is removed and inserted between adjacent nodes 3 and 4.

2.3.5.7 Swap(1,1)   In this structure, the positions of any node pair located in different routes are exchanged. In Figure 2.4b, the positions of nodes 4 and 8 are exchanged.

2.3.5.8 Shift(2,0)   This structure removes an adjacent node pair and transfers them to another route. In Figure 2.4c, adjacent nodes 5 and 6 are removed and inserted between nodes 4 and 0.

2.3.5.9 Swap(2,1)   This structure exchanges the positions of two adjacent customers with a customer from a different route. In Figure 2.4d, adjacent nodes 7 and 8 are exchanged with node 4.

2.3.5.10 Swap(2,2)   This structure exchanges the positions of two adjacent node pairs from different routes. In Figure 2.4e, adjacent nodes 1 and 2 are exchanged with adjacent nodes 8 and 9.

2.3.6 Applied VND Algorithm (p_VND)

VND algorithm is employed at both stages of the algorithm to intensify the search process. The basic steps of the algorithm are illustrated in Figure 2.5. We utilize five inter-route neighborhood structures, Swap(1,1), Shift(1,0), Swap(2,1), Shift(2,0), and Swap(2,2), for VND. Whenever a new improved solution is obtained, the procedure VND_intra is implemented on the routes modified by the current inter-route structure to further improve the quality of the solution. When the procedure is called, four intra-route neighborhood structures, General Swap, 2-opt, Single Insertion, and Block Insertion, are exhaustively applied to the modified routes. The strategy of using intra-route structures after obtaining an improvement by implementing inter-route ones can be seen in Subramanian et al. (2010) and Goksal et al. (2013). The order of the neighborhood structures that influences the performance of the algorithm is determined based on

```
Procedure p_VND
Input: x, inter-route neighborhood structures NS_inter
Output: x
k←1
while k≤kₘₐₓ
    find the best neighbor (x') of x with respect to the kᵗʰ
    neighborhood structure;
    if f(x')<f(x)
        x''←VND_intra(x', NS_intra);
        x←x'';
        k←1;
    else
        k←k+1;
    end
end
```

Figure 2.5   Main steps of the applied VND algorithm.

our preliminary experiments. The order of the inter-route structures is determined as follows: Swap(1,1) is used as the first structure, and Shift(1,0), Shift(2,0), Swap(2,1), and Swap(2,2) are implemented after Swap(1,1), respectively.

2.4 Numerical Study

The proposed algorithm is coded in MATLAB® 7.8.0 and executed on a 3.00 GHz Pentium 4 computer. In order to test the performance of the algorithm for the VRPSPD, the benchmark instances presented by Dethloff (2001) are used. The dataset that consists of 40 problems with 50 customers is classified according to two different geographical scenarios and labeled as either SCA or CON. In SCA instances, the coordinate values of the customers are uniformly distributed within [0, 100]. In CON datasets, half of the customers are scattered in a similar way with SCA, while the coordinates of the other half are uniformly distributed in [100/3, 200/3]. Thus, the scenario CON represents urban areas where a big portion of the population is located in small regions. The delivery amount of customer i (d_i) is distributed in [0, 100]. The pickup amount of each customer is determined by the equation of $p_i = (0.5 + r_i) \cdot d_i$, where r_i is a random number uniformly distributed over the interval [0, 1]. The vehicle capacities are generated from the equation of $C = \sum_{i \in J} d_i / \mu$, where μ represents the minimum number of vehicles required. In the instance descriptor, the digit after the letters for the geographical scenario in Table 2.1 represents the respective value of μ, which is chosen to be 3 or 8 (Dethloff 2001).

Table 2.1 Computational Results for the VRPSPD Benchmark Instances of Dethloff (2001)

INSTANCE	N	R&P BEST SOL.	G&A BEST SOL.	ZTK BEST SOL.	SDBOF BEST SOL.	GKA BEST SOL.	HYBRID_SALS BEST SOL.	HYBRID_SALS AVG. SOL.	HYBRID_SALS DEV. OF BEST SOL.
SCA3-0	50	636.1	635.62	635.62	635.62	635.62	636.06	639.03	0.0006
SCA3-1	50	697.8	697.84	697.84	697.84	697.84	697.84	703.13	0.0000
SCA3-2	50	659.3	659.34	659.34	659.34	659.34	659.34	662.02	0.0000
SCA3-3	50	680.6	680.04	680.04	680.04	680.04	680.04	682.20	0.0000
SCA3-4	50	690.5	690.50	690.50	690.50	690.50	690.50	693.46	0.0000
SCA3-5	50	659.9	659.90	659.90	659.90	659.90	659.90	663.02	0.0000
SCA3-6	50	651.1	651.09	651.09	651.09	651.09	651.09	655.66	0.0000
SCA3-7	50	666.1	659.17	659.17	659.17	659.17	659.17	667.20	0.0000
SCA3-8	50	719.5	719.47	719.47	719.47	719.47	719.47	724.46	0.0000
SCA3-9	50	681.0	681.00	681.00	681.00	681.00	681.00	689.73	0.0000
SCA8-0	50	975.1	961.50	961.50	961.50	961.50	961.50	967.69	0.0000
SCA8-1	50	1052.4	1049.65	1049.65	1049.65	1049.65	1049.65	1051.02	0.0000
SCA8-2	50	1044.5	1042.69	1039.64	1039.64	1039.64	1039.64	1047.80	0.0000
SCA8-3	50	999.1	983.34	983.34	983.34	983.34	983.34	989.99	0.0000
SCA8-4	50	1065.5	1065.49	1065.49	1065.49	1065.49	1065.49	1070.66	0.0000
SCA8-5	50	1027.1	1027.08	1027.08	1027.08	1027.08	1027.08	1031.21	0.0000
SCA8-6	50	977.0	971.82	971.82	971.82	971.82	971.82	975.81	0.0000
SCA8-7	50	1061.0	1052.17	1051.28	1051.28	1051.28	1051.28	1058.56	0.0000
SCA8-8	50	1071.2	1071.18	1071.18	1071.18	1071.18	1071.18	1078.30	0.0000
SCA8-9	50	1060.5	1060.50	1060.50	1060.50	1060.50	1060.50	1069.10	0.0000
CON3-0	50	616.5	616.52	616.52	616.52	616.52	616.52	621.67	0.0000

(Continued)

Table 2.1 (*Continued*) Computational Results for the VRPSPD Benchmark Instances of Dethloff (2001)

INSTANCE	N	R&P BEST SOL.	G&A BEST SOL.	ZTK BEST SOL.	SDBOF BEST SOL.	GKA BEST SOL.	HYBRID_SALS BEST SOL.	HYBRID_SALS AVG. SOL.	HYBRID_SALS DEV. OF BEST SOL.
CON3-2	50	521.4	**518.00**	**518.00**	**518.00**	**518.00**	520.10	521.01	0.0040
CON3-3	50	**591.2**	**591.19**	**591.19**	**591.19**	**591.19**	**591.19**	591.66	0.0000
CON3-4	50	**588.8**	**588.79**	**588.79**	**588.79**	**588.79**	**588.79**	593.24	0.0000
CON3-5	50	**563.7**	**563.70**	**563.70**	**563.70**	**563.70**	**563.70**	566.01	0.0000
CON3-6	50	500.8	**499.05**	**499.05**	**499.05**	**499.05**	**499.05**	505.81	0.0000
CON3-7	50	**576.5**	**576.48**	**576.48**	**576.48**	**576.48**	**576.48**	580.70	0.0000
CON3-8	50	**523.1**	**523.05**	**523.05**	**523.05**	**523.05**	**523.05**	523.05	0.0000
CON3-9	50	586.4	**578.25**	**578.25**	**578.25**	**578.25**	**578.25**	580.15	0.0000
CON8-0	50	**857.2**	**857.17**	**857.17**	**857.17**	**857.17**	**857.17**	857.99	0.0000
CON8-1	50	**740.9**	**740.85**	**740.85**	**740.85**	**740.85**	**740.85**	747.82	0.0000
CON8-2	50	716.0	**712.89**	**712.89**	**712.89**	**712.89**	**712.89**	716.64	0.0000
CON8-3	50	**811.1**	**811.07**	**811.07**	**811.07**	**811.07**	**811.07**	812.17	0.0000
CON8-4	50	**772.3**	**772.25**	**772.25**	**772.25**	**772.25**	**772.25**	778.92	0.0000
CON8-5	50	755.7	**754.88**	**754.88**	**754.88**	**754.88**	**754.88**	755.49	0.0000
CON8-6	50	693.1	**678.92**	**678.92**	**678.92**	**678.92**	**678.92**	687.66	0.0000
CON8-7	50	814.8	**811.96**	**811.96**	**811.96**	**811.96**	**811.96**	812.72	0.0000
CON8-8	50	774.0	**767.53**	**767.53**	**767.53**	**767.53**	**767.53**	768.79	0.0000
CON8-9	50	809.3	**809.00**	**809.00**	**809.00**	**809.00**	**809.00**	809.66	0.0000
BKS f.		22	38	40	40	40	38		
Avg. dev.		0.0032	0.0001	0.0000	0.0000	0.0000	0.0001		0.0000

Note: Bold numbers indicate that the algorithm has reached the best solution.

2.4.1 Computational Results

The proposed hybrid_SALS is run 10 times for each benchmark problem. A statistical analysis for the dataset is also conducted by applying a paired-t test at a significance level of $\alpha = 0.05$ to reveal whether there exists significant differences between hybrid_SALS and other methods in terms of solution quality or not. Thus, we set up the null hypothesis as $H_0 = \mu_{\text{hybrid\_SALS}} - \mu_{CA} = 0$ and a two-sided alternative hypothesis as $H_1 = \mu_{\text{hybrid\_SALS}} - \mu_{CA} \neq 0$. The symbols $\mu_{\text{hybrid\_SALS}}$ and μ_{CA} represent the population mean for hybrid_SALS and the compared algorithm, respectively.

We terminate the algorithm after 1000 successive iterations without any improvement. We compare our algorithm with the following best-known algorithms available in the literature:

- R&P: LNS (Ropke and Pisinger 2006)
- G&A: Ant colony system (Gajpal and Abad 2009)
- SDBOF: Parallel ILS (Subramanian et al. 2010)
- ZTK: Adaptive memory methodology (Zachariadis et al. 2010)
- GKA: Hybrid discrete PSO (Goksal et al. 2013)

Table 2.1 shows the computational results of the proposed algorithm for the VRPSPD instances. For hybrid_SALS, we report the best solution and the average solution observed from each instance. For other algorithms, only the best found solutions are reported because the average results are not reported in all studies. Additionally, the number of best-known solutions (BKS) found and the average deviations from BKS (avg. dev.) are reported in last two rows. Bold numbers indicate that the algorithm has reached the best solution. As seen from the table, hybrid_SALS produces 38 BKS out of 40 problem instances. While hybrid_SALS fails to find the best solutions for two instances, the solutions obtained from these instances are so close to the best results that the average deviation of obtained solutions from the best results is 0.0001. Results of the paired-t test for the corresponding dataset are presented in Table 2.2. As seen from the table, while hybrid_SALS is statistically significantly different from R&P, there is no statistically significant difference between hybrid_SALS and other algorithms. This outcome shows that our proposed algorithm performs as good as the most sophisticated methods proposed for the VRPSPD.

Table 2.2 Results of Paired-t Test for the VRPSPD Benchmark Instances of Dethloff (2001)

PAIRS (HYBRID_SALS—COMPARED ALGORITHM)	MEAN DIFFERENCE	P-VALUE
hybrid_SALS—GKA	0.063	0.241
hybrid_SALS—SDBOF	0.063	0.241
hybrid_SALS—ZTK	0.063	0.241
hybrid_SALS—G&A	−0.035	0.720
hybrid_SALS—R&P	−2.465	0.001

A fair comparison of the solution methods in terms of the computational time cannot be made because of different influencing factors such as software, hardware, and coding. However, our computational experiments indicate that the proposed hybrid_SALS provides its ultimate solution in less than 3 min for all instances. This result indicates that the developed algorithm is effective in solving the VRPSPD in reasonable computation time.

2.5 Conclusion

In this study, we deal with a variant of the classical VRP that is the VRPSPD arising mainly in reverse logistic systems that include both distribution and collection operations. The VRPSPD has gained increased importance in recent years since the integration of pickup and delivery operations has become a vital task for companies to minimize costs.

In order to solve the problem, we have developed a local search algorithm (hybrid_SALS) that is constructed by hybridizing an SA-inspired algorithm with VND. The applied SA-inspired algorithm, SALS, is a parameter-free metaheuristic whose nonmonotonic threshold function provides the required diversification to the search. On the other hand, the VND algorithm, which is a well-known local search method, has been implemented to intensify the search process. Since hybrid_SALS has been developed using parameter-free algorithms, it does not need any parameter setting, which consequently makes its implementation to the optimization problems much easier.

The performance of hybrid_SALS is tested by using well-known benchmark instances for the VRPSPD. The computational results indicate that the hybrid_SALS generates high-quality solutions for the instances and performs as well as the most sophisticated methods proposed for the VRPSPD up to date. Moreover, given its simplicity, hybrid_SALS has the advantage of being a parameter-free algorithm in comparison with the existing algorithms.

References

Ai, T. J. and V. Kachitvichyanukul. 2009. A particle swarm optimization for the vehicle routing problem with simultaneous pickup and delivery. *Computers & Operations Research* 36 (5):1693–1702.

Alabas-Uslu, C. and B. Dengiz. 2011. A self-adaptive local search algorithm for the classical vehicle routing problem. *Expert Systems with Applications* 38 (7):8990–8998.

Bianchessi, N. and G. Righini. 2007. Heuristic algorithms for the vehicle routing problem with simultaneous pick-up and delivery. *Computers & Operations Research* 34 (2):578–594.

Çatay, B. 2010. A new saving-based ant algorithm for the vehicle routing problem with simultaneous pickup and delivery. *Expert Systems with Applications* 37 (10):6809–6817.

Chen, J. F. and T. H. Wu. 2005. Vehicle routing problem with simultaneous deliveries and pickups. *Journal of the Operational Research Society* 57 (5):579–587.

Crispim, J. and J. Brandão. 2005. Metaheuristics applied to mixed and simultaneous extensions of vehicle routing problems with backhauls. *Journal of the Operational Research Society* 56 (11):1296–1302.

Dell'Amico, M., G. Righini, and M. Salani. 2006. A branch-and-price approach to the vehicle routing problem with simultaneous distribution and collection. *Transportation Science* 40 (2):235–247.

Dethloff, J. 2001. Vehicle routing and reverse logistics: The vehicle routing problem with simultaneous delivery and pick-up. *OR-Spektrum* 23 (1):79–96.

Gajpal, Y. and P. Abad. 2009. An ant colony system (ACS) for vehicle routing problem with simultaneous delivery and pickup. *Computers & Operations Research* 36 (12):3215–3223.

Gajpal, Y. and P. Abad. 2010. Saving-based algorithms for vehicle routing problem with simultaneous pickup and delivery. *Journal of the Operational Research Society* 61 (10):1498–1509.

Goksal, F. P., I. Karaoglan, and F. Altiparmak. 2013. A hybrid discrete particle swarm optimization for vehicle routing problem with simultaneous pickup and delivery. *Computers & Industrial Engineering* 65 (1):39–53.

Min, H. 1989. The multiple vehicle routing problem with simultaneous delivery and pick-up points. *Transportation Research Part A: General* 23 (5):377–386.

Mladenović, N. and P. Hansen. 1997. Variable neighborhood search. *Computers & Operations Research* 24 (11):1097–1100.

Montané, F. A. T. and R. D. Galvão. 2006. A tabu search algorithm for the vehicle routing problem with simultaneous pick-up and delivery service. *Computers & Operations Research* 33 (3):595–619.

Nagy, G. and S. Salhi. 2005. Heuristic algorithms for single and multiple depot vehicle routing problems with pickups and deliveries. *European Journal of Operational Research* 162 (1):126–141.

Prins, C. 2004. A simple and effective evolutionary algorithm for the vehicle routing problem. *Computers & Operations Research* 31 (12):1985–2002.

Ropke, S. and D. Pisinger. 2006. A unified heuristic for a large class of vehicle routing problems with backhauls. *European Journal of Operational Research* 171 (3):750–775.

Subramanian, A., L. M. A. Drummond, C. Bentes, L. S. Ochi, and R. Farias. 2010. A parallel heuristic for the vehicle routing problem with simultaneous pickup and delivery. *Computers & Operations Research* 37 (11): 1899–1911.

Subramanian, A., E. Uchoa, A. A. Pessoa, and L. S. Ochi. 2011. Branch-and-cut with lazy separation for the vehicle routing problem with simultaneous pickup and delivery. *Operations Research Letters* 39 (5):338–341.

Tasan, A. S. and M. Gen. 2012. A genetic algorithm based approach to vehicle routing problem with simultaneous pick-up and deliveries. *Computers & Industrial Engineering* 62 (3):755–761.

Wassan, N. A., A. H. Wassan, and G. Nagy. 2008. A reactive tabu search algorithm for the vehicle routing problem with simultaneous pickups and deliveries. *Journal of Combinatorial Optimization* 15 (4):368–386.

Zachariadis, E. E. and C. T. Kiranoudis. 2011. A local search metaheuristic algorithm for the vehicle routing problem with simultaneous pick-ups and deliveries. *Expert Systems with Applications* 38 (3):2717–2726.

Zachariadis, E. E., C. D. Tarantilis, and C. T. Kiranoudis. 2009. A hybrid metaheuristic algorithm for the vehicle routing problem with simultaneous delivery and pick-up service. *Expert Systems with Applications* 36 (2): 1070–1081.

Zachariadis, E. E., C. D. Tarantilis, and C. T. Kiranoudis. 2010. An adaptive memory methodology for the vehicle routing problem with simultaneous pick-ups and deliveries. *European Journal of Operational Research* 202 (2): 401–411.

3

OPTIMAL FENCING IN AIRLINE INDUSTRY WITH DEMAND LEAKAGE

SYED ASIF RAZA AND MIHAELA TURIAC

Contents

3.1 Introduction

Four decades from its beginning in the airline industry, revenue management (RM) practice evolved rapidly to complex systems with applicability in many industries and gained researchers' attention. McGill and Ryzin (1999) and more recently Chiang et al. (2007) presented in detail an overview of RM research. After the 1983 US Airline Deregulation Act, considered as one of the most important applications of management science and operations research (Bell, 1998), two essential features remained in RM practice: demand segmentation (which for an airline means managing the set of fare classes) and fare classes availability management. In particular, airline RM research has focused on four categories: forecasting, overbooking, quantity/inventory (booking) control and pricing; however, as noticed in Cote et al. (2003), integration of pricing and quantity controls is expected to improve the firm's revenue. The tactics and strategies of RM are applied in general in business that has a fixed or perishable resource like the flight seats in airline business or the hotels' rooms.

As noticed in Anon (n.d.), the general RM practices are classified into quantity-based RM and price-based RM. A quantity-based RM problem is designed as a revenue optimization model in which the resource allocation can be adjusted efficiently for predetermined prices. This practice is well applied in airline industry and is usually addressed as seat inventory control problem. The price-based RM is applied to maximize the revenue by optimizing the pricing when the available resource is fixed. Such typical business frame is commonly observed in retail industry where the simplest form of RM has been identified in the Newsvendor (Newsboy) problem, considered by Petruzzi and Dada (1999) as a building block in stochastic inventory control and an excellent tool for examining how operational and marketing issues interact in the decision-making process.

Pricing is one of the main cores of yield management, also known as RM practice. The fundamental concept is to segment the market into multiple market segments using a differentiation price, which will offer potentially a different price or sale condition. One real-life example of price differentiation practice can be observed in the sale tickets offered by airlines to passengers who are willing to pay in advance and accept penalties for changing or canceling tickets, while for the late-arriving passengers who are less price sensitive and more willing to pay for their tickets less restrictive, the airline reserves part of its capacity. In airline RM, this tactic is usually referred to as fare price differentiation and is among the principal strategy used to segment the demand from one fare class to multiple fare classes. As RM tactic, pricing is applied also by hotels that often set higher prices for weekdays' room rates expected to be reserved by business customers, compared to weekend rates, which are more desirable for leisure customers. Similar to airlines, hotels do apply several penalties such as cancellation restrictions, fees for changes in reservation, or nonrefundable lower-priced room rates to achieve buy-down. Another example of how price differentiation which leads customers to different channels is the online versus retail store sales, where a firm may offer discounted prices for online sales with no option of touch and feel, and higher prices for retail store sale due to the option of interacting with sales staff or with products.

Integration of pricing and seat allocation in airline RM began with Weatherford (1997), who considered normally distributed customers' demand with mean as a linear function of price. Feng and Xiao (2001) studied the integration of capacity and pricing decisions for perishable assets in a comprehensive model with stochastic demand. Cote et al. (2003) developed a bilevel mathematical programming approach for joint determination of fare price and seat allocation. Raza and Akgunduz (2008) proposed a game theoretic model for an integrated approach of fare pricing duopoly competition with seat allocation and extended their work in Raza and Akgunduz (2010) to a cooperative game setting using bargain solution.

Many research studies (see Anon, n.d.; Philips, 2005) reported that market segmentation from price differentiation augments profitability; however, different prices for distinct market segments often cause customers' cannibalization, referred also as demand leakage from one market segment to another. The effects of market segmentation with demand leakage on a firm's pricing and inventory decisions were studied in Zhang and Bell (2007). To mitigate cannibalization and maintain the fences that differentiate the market segments, one common practice is to improve the fences by introducing restrictions that would prevent customers from migrating between market segments. A fence can be referred as a device designed to preserve the market segments formed after price differentiation. Among such devices commonly observed are early purchase, prolong processing time, return penalties, channel of purchase, etc. An overview and taxonomy of price fencing in RM practice can be found in Zhang and Bell (2010). Li (2001) investigated pricing of nonstorable perishable goods in a deterministic demand case with imperfect market segmentation and purchase restriction with an application to airline fare pricing. The interest on fencing led earlier researchers to identify that maintaining appropriate fences is very essential for an efficient RM (see Hanks et al., 2002; Kimes, 2002; Zhang et al., 2010). However, there are still many concerns on fencing as a business practice especially in the context of airline industry, such as How can an airline control demand leakage through investment in fencing, and what will be the optimal investment? How profitable is to integrate the pricing, seat inventory control, and fencing investment decisions for an airline?

A related study was conducted by Zhang et al. (2010) for an uncapacitated pricing and fencing investment decision problem of a firm. Noticeably, airline RM modeling is significantly different from a typical firm; however, both problems can resemble newsvendor problem (see Philips [2005] for details on newsvendor problem and RM relationship). In airline RM, the demand arrival is assumed sequential, and therefore, the lower fare price class demand is observed prior to the respective higher fare class. In response to this, the airline exercises a nested control that reserves certain seats for passengers willing to pay a higher fare price and arrive later to purchase tickets. This control is referred in airline RM literature as nested booking control (McGill and Ryzin, 1999; Chiang et al., 2007). Furthermore, in airline RM, unlike the uncapacitated firm's problem, there is a limited capacity represented by the cabin seats. Lastly, in airline RM, the costs incurred in relation to seat inventory and related flight services are often ignored in most of the airline RM models, with no exception in this study. Thus, the focus of this study is to revisit the problem of RM with demand leakages and fencing investments in the airline context. We first present the model for an airline RM with no fencing investment to mitigate the demand leakage, and then we extend the problem with fencing improvement decision for the airline to mitigate or augment the demand leakage through additional investment. Later, the models are analyzed, and the optimal fare pricing, seat inventory control (nested booking control), and fencing decisions are determined. Finally, a numerical experimentation study is presented to highlight the impact of some significant problem-related factors such as demand variability and leakage rate onto the airline's RM decision. Additionally, the fencing investment decision is also studied numerically to determine the airline's decision toward demand leakage control.

3.2 Model Development

We propose a single-leg RM model for an airline that exercises an optimal integrated control on fare pricing, seat inventory control, and fencing investment. The airline activates in monopoly and segments the market into two segments using differentiated prices strategy. The market segmentation is assumed imperfect; thus, the customers cannibalize from the full fare class to discounted fare class. In order to mitigate demand leakage, a fencing investment is proposed to improve the airline's market fences. Assuming stochastic demand, the airline performs a nested control over the single-resource capacity following Littlewood's (1972) rule for customers' sequential arrival. Let c denote the capacity of the airplane cabin. The airline offers seats in the cabin for two adjacent fare classes: class 1, designated for business travelers willing to pay full fare price p_1, and class 2, for leisure travelers willing to pay a discounted fare price p_2, where $p_1 \geq p_2$. Like many other studies from RM literature (see Choi, 1996; Chiang and Monahan, 2005; Zhang et al., 2010), we assume a linear price-dependent demand, which, in a riskless perfect market segmentation case, is given by $[\alpha_i - \beta_i p_i]^+$, where $\alpha_i, \beta_i > 0$, $\forall i = \{1,2\}$. After the price differentiation strategy, the market segments created are assumed imperfect, and the airline observes γ proportion of passengers cannibalizing from full fare to discounted fare class. To model this behavior, we use a liner function given by $\gamma(p_1 - p_2)$, where $\gamma \geq 0$ represents leakage rate. If $\gamma = 0$, then the airline is considered to have a perfect fence. Thus, the deterministic linear demand curves influenced by demand leakage would be

$$y_1(p_1, p_2, \gamma) = \alpha_1 - \beta_1 p_1 - \gamma(p_1 - p_2) \tag{3.1}$$

$$y_2(p_1, p_2, \gamma) = \alpha_2 - \beta_2 p_2 + \gamma(p_1 - p_2) \tag{3.2}$$

The stochastic demand D_i, for fare class i, $\forall i = \{1,2\}$, is modeled from deterministic demand y_i and a random factor ξ_i, where ξ_i has price-independent probability distribution $f_i(\xi_i)$ and cumulative probability distribution $F_i(\xi_i)$, both continuous, twice differentiable, invertible, and following an increasing failure rate. Moreover, ξ_i is assumed in $[\underline{\xi}_i, \overline{\xi}_i]$ with mean μ_i and standard deviation σ_i. This study follows Mostard et al. (2005); in this study, we have assumed $\xi_i \in [-\sqrt{3}\sigma, \sqrt{3}\sigma]$. Following Petruzzi and Dada (1999), an additive approach is assumed for D_i, $\forall i = \{1,2\}$, such that

$$D_i\left(y_i, \xi_i\right) = y_i + \xi_i, \quad \forall i = \{1,2\} \tag{3.3}$$

A list of notations relevant to the model is presented in Table 3.1.

Table 3.1 Model Parameters and Notations

PARAMETERS	
c	Inventory capacity
α_i	Maximum perceived demand in fare class i, $\forall i = \{1,2\}$
β_i	Price sensitivity of deterministic demand in fare class i, $\forall i = \{1,2\}$
$y_i = y_i(p_1,p_2,\gamma)$	Deterministic demand in fare class i, $\forall i = \{1,2\}$
$\xi_i \in [\underline{\xi}_i, \overline{\xi}_i]$	Stochastic demand factor for fare class i, $\forall i = \{1,2\}$
$f_i(\xi_i)$	Probability distribution function of stochastic factor ξ_i, $\forall i = \{1,2\}$
$F_i(\xi_i)$	Cumulative probability distribution of stochastic factor ξ_i, $\forall i = \{1,2\}$
$D_i = D_i(p_1,p_2,\gamma,\xi_i)$	Price-dependent stochastic demand in fare class i, $\forall i = \{1,2\}$
$\hat{\pi}$	Revenue without fencing investment
$E(\pi)$	Revenue with fencing investment
$G(\gamma)$	Cost of fencing
G_0	Initial cost of fencing
$*$	Optimal of a decision control parameter
DECISION VARIABLES	
p_i	Price in fare class i, $\forall i = \{1,2\}$
x_i	Capacity allocation for fare class i, $\forall i = \{1,2\}$
γ	Demand leakage factor, $0 \leq \gamma \leq \overline{\gamma}$

3.2.1 No Fencing Investment

As specified earlier, D_i $\forall i = \{1,2\}$ is assumed with sequential arrivals so that the airline observes discounted fare class demand prior to the full fare class demand. Thus, the airline's revenue from offering two fare classes to its passengers while performing a nested control of the inventory capacity would be given as:

$$\hat{\pi} = p_1 \min\{x_1 + \min\{x_2, D_2\}, D_1\} + p_2 \min\{x_2, D_2\} \qquad (3.4)$$

The revenue function from this equation can be simplified as in the following equation (see Appendix):

$$\hat{\pi} = p_1 x_1 + p_2 x_2 + (p_1 - p_2) \int_{\underline{\xi}_2}^{x_2 - y_2} F_2(\xi_2)\, d\xi_2$$

$$- p_1 \int_{\underline{\xi}_1}^{x_1 + \int_{\underline{\xi}_2}^{x_2 - y_2} F_2(\xi_2) - y_1} F_1(\xi_1)\, d\xi_1 \qquad (3.5)$$

The first two terms in Equation 3.5 represent the deterministic riskless profit; the third term is the expression of revenue gain from nested capacity control mainly due to price differential $(p_1 > p_2)$, where expected demand $\int_{\xi_2}^{x_2 - y_2} F_2(\xi_2)d\xi_2$ is protected from discounted fare class 2 and reserved for full fare class 1. The last term represents the loss in revenue due to an observed demand for full fare class, which is lower than the actual capacity allocated $x_1 + \int_{\xi_2}^{x_2 - y_2} F_2(\xi_2)d\xi_2$. The airline problem in this case is formulated as follows:

$$P: \quad \underset{p_1, p_2, x_1, x_2}{\text{Max}} \quad \hat{\pi} \tag{3.6}$$

$$\text{subject to}: x_1 + x_2 \leq c \tag{3.7}$$

3.2.2 With Fencing Investment

Given that the price differentiation strategy results in imperfect fences and hence, in demand leakage, the airline's problem extends to diminishing the customers' shifting from full fare class to discounted fare class. Without loss of generality, we presume that the airline decides to increase fencing levels through an investment of specific costs. Suppose that for reaching γ leakage, the airline must bear a cost, $G(\gamma)$, assumed nonnegative, continuous and monotonically decreasing in γ. Thus, the revenue function from Equation 3.5 is adjusted by the fencing cost $G(\gamma)$, and the airline problem is formulated now as a constraint nonlinear optimization problem, P':

$$P': \quad \underset{p_1, p_2, x_1, x_2, \gamma}{\text{Max}} \quad \pi = \hat{\pi} - G(\gamma) \tag{3.8}$$

$$\text{subject to}: x_1 + x_2 \leq c \tag{3.9}$$

The optimal expected revenue when fencing investment decisions are taken would be $\pi^*(p_1^*, p_2^*, x_1^*, x_2^*, \gamma^*)$, and the airline's problem is to determine the optimal integrated decisions on fare prices p_1^* and p_2^*, seat inventory control x_1^* and x_2^* and investment $G(\gamma^*)$ for demand leakage γ^*. It is important to notice here that the optimality of revenue, $\hat{\pi}$, from P would be an upper bound on the optimal total expected revenue, π, from P', when the airline decides on fencing investment.

3.3 Model Analysis

We address first the airline's optimization problem, P, to jointly determine the fare pricing and seat inventory control. Due to computational complexity in structural properties analysis of the revenue function, we provide two approaches to solve the model: sequential (hierarchical) optimization and joint optimization. In hierarchical optimization, the decision control parameters are optimized sequentially such that the airline determines first the optimal fare prices, p_1^* and p_2^*, and later, the optimal inventory control decisions, x_1^* and x_2^*. In problem P', an additional decision parameter γ is considered to determine the fencing investment, achieved also by sequential optimization. This approach of addressing inventory control and pricing decisions has been applied in several studies (see Smith et al., 2007; Zhang et al. 2010).

3.3.1 Hierarchical Optimization

To apply the sequential approach, we consider problem P, and we use a hierarchical optimization procedure while demand leakage rate, γ, is fixed, thus, no investment assumed to control the fencing via demand leakage rate. In our pursuit to determine the fare pricing, while ignoring the seat inventory control decisions x_1 and x_2, and since pricing decisions are mostly dependent on the price-dependent deterministic demands, y_i, $\forall i = \{1,2\}$, we formulate a deterministic version of problem, P, as problem, DP. Given that demand uncertainties are ignored, the stochastic demands D_1 and D_2 are approximated with the expectations, $z_1 = y_1 + \mu_1$ and $z_2 = y_2 + \mu_2$, respectively. Thus, the deterministic problem, DP, of the airline would be

$$DP : \underset{p_1, p_2}{\text{Max }} \pi^d = p_1 z_1 + p_2 z_2 \qquad (3.10)$$

$$\text{subject to} : z_1 + z_2 \leq c \qquad (3.11)$$

For DP, we can determine the optimal fare prices, p_1^* and p_2^*, as outlined in Proposition 3.1.

Proposition 3.1   In DP, the following holds:

1. The optimal prices p_i^*, $\forall i = \{1,2\}$ are determined by solving the following system of nonlinear equations:

$$\alpha_1 - 2p_1(\beta_1 + \gamma) + 2p_2\gamma + \lambda\beta_1 + \mu_1 = 0 \qquad (3.12)$$

$$\alpha_2 - 2p_2(\beta_2 + \gamma) + 2p_1\gamma + \lambda\beta_2 + \mu_2 = 0 \qquad (3.13)$$

$$c - (\alpha_1 + \alpha_2) + \beta_1 p_1 + \beta_2 p_2 - (\mu_1 + \mu_2) = 0 \qquad (3.14)$$

2. π^d is jointly concave in p_i, $\forall i = \{1,2\}$.

Proof: See Appendix.

Now, we reconsider problem P with no fencing investment while the stochastic demand assumption with sequential arrivals holds. We create the stochastic problem for the airline as:

$$P: \underset{p_1,p_2,x_1,x_2}{\text{Max}} \quad \hat{\pi} = p_1x_1 + p_2x_2 + (p_1 - p_2) \int_{\xi_2}^{x_2 - y_2} F_2(\xi_2)d\xi_2$$

$$-p_1 \int_{\xi_1}^{x_1 + \int_{\xi_2}^{x_2-y_2} F_2(\xi_2) - y_1} F_1(\xi_1)d\xi_1 \qquad (3.15)$$

$$\text{subject to}: x_1 + x_2 \leq c \qquad (3.16)$$

In P, the optimal expected revenue would be given by $\hat{\pi}^*(p_1^*, p_2^*, x_1^*, x_2^*)$, where p_1^* and p_2^* are the optimal fare prices, and x_1^* and x_2^*, are the optimal seat inventory controls of full fare and discounted fare class, respectively. The constraint in Equation 3.16 is the flight cabin limitation. In this problem, the optimal fare prices, p_1^* and p_2^*, are first obtained from Proposition 3.1, and the expected total revenue function, $\hat{\pi}$, in Equation 3.15 can be optimized to determine the optimal seat inventory controls x_1^* and x_2^*, as outlined in Proposition 3.2.

Proposition 3.2 In problem P, the following holds:

1. Given that the optimal fare prices p_1 and p_2 are fixed, the optimal booking limit x_2^* is such that $x_2^* = y_2 + \sqrt{3}\,\sigma$.

2. $\hat{\pi}$ is jointly concave in booking limits x_1 and x_2 if $p_1\Phi_1 -$
$(p_1-p_2)\geq 0$, where $\Phi_1 = F_1\left(x_1 + \int_{\xi_2}^{x_2-y_2} F_2(\xi_2) - y_1\right)$ and $\Phi_2 =$
$F_2(x_2-y_2)$.

Proof: See Appendix.

3.3.2 Joint Optimization

In this section, we extend the optimization procedure approached earlier for problem *P*. Proposition 3.3 outlines the procedure to determine the joint optimal control for problem *P*. The decision controls here are the optimal fare prices p_1^* and p_2^* and the optimal seat inventory controls x_1^* and x_2^* Due to the complex structure of the revenue function, $\hat{\pi}$, mainly contributed from demand uncertainty, sequential demand arrival (nested control), and price-dependent demand leakage, proving the joint concavity in all decision variable could be a prohibitive task and it is not explored in this study. However, joint concavity of $\hat{\pi}$ is shown in seat inventory control for fixed fare prices and vice versa. These results may be found more restrictive in terms of a more general condition for joint concavity of $\hat{\pi}$, but again, given that the complex structure of the revenue function, an analytical framework to derive a less-restrictive condition seems limiting.

Proposition 3.3 In problem *P'*, the following holds for the joint optimization:

1. For a fixed set of inventory control x_i, $\forall i = \{1,2\}$, π is jointly concave in fare prices p_i, $\forall i = \{1,2\}$, as long as $p_1\phi_1 t_1 t_2 - (\beta_2 + 2\gamma + \gamma(\beta_2 + \gamma)t_3) \geq 0$, where $t_1 = \beta_1 + \gamma(1-\Phi_2)$, $t_2 = \Phi_2\beta_2 - \gamma(1-\Phi_2)$, $t_3 = p_1\Phi_1 - (p_1-p_2)$, and $t_1, t_2, t_3 \geq 0$.

2. The optimal fare prices p_1^* and p_2^*, seat inventory controls x_1^* and x_2^*, and demand leakage γ^* are determined by solving the following system of nonlinear equations:

$$p_1(1-\Phi_1) - p_2 + \Phi_2\left(p_1\Phi_1 - (p_1-p_2)\right) = 0 \qquad (3.17)$$

$$x_1 - I_1 + I_2 - (p_1-p_2)\Phi_2\gamma - p_1\Phi_1(\beta_1 + \gamma(1-\Phi_2)) = 0 \quad (3.18)$$

$$x_2 - I_2 + (p_1 - p_2)\Phi_{22}(\beta_2 + \gamma) - p_1\Phi_1(\beta_2\Phi_2 - \gamma(1 - \Phi_2)) = 0$$
$$(3.19)$$

$$-\Phi_2(p_1 - p_2)^2 - p_1\Phi_1(p_1 - p_2)(1 - \Phi_2) - \frac{\partial G}{\partial \gamma} = 0 \quad (3.20)$$

$$c - x_1 - x_2 = 0 \qquad (3.21)$$

Next, we study problem P', which includes also the cost of fencing. A similar study in a firm's context with no capacity constraints has been reported in Zhang et al. (2010). There are two types of fencing cost models considered here: linear fencing cost and nonlinear fencing cost.

3.3.2.1 Linear Fencing Cost For a linear fencing cost approach, the cost function of the fencing investment is linearly linked to the leakage rate, γ. We define the linear cost function considering the range of leakage as $G(\gamma) = G_0 - (G_0/K)\gamma$, where $G_0 > 0$ is the cost of null leakage, when the perfect fence is achieved ($\gamma = 0$), and $K > 0$ is the maximum leakage level when there is no initiative to invest in fencing and $G(\gamma) = 0$.

Proposition 3.4 Given, x_i, p_i, $\forall i = \{1,2\}$ and a linear fencing cost, $G(\gamma) = G_0 - (G_0/K)\gamma$, the following hold in the problem, P':

1. The revenue, π is quasi-concave (unimodal) in γ, if $\phi_2(p_1 - p_2) - p_1(\phi_1(1 - \Phi_2)^2 + \Phi_1(1 - \phi_2)) \le 0$.
2. The optimal leakage rate, γ^* can be determined by solving,
$$(p_1 - p_2)(\Phi_2(p_1 - p_2) + p_1\Phi_1(1 - \Phi_2)) + \frac{G_0}{K} = 0.$$

Proof: See Appendix.

3.3.2.2 Nonlinear Fencing Cost In the case of nonlinear fencing cost, it is assumed that for a small leakage rate the cost of fence grows rapidly and then slowly when leakage rate reaches high levels. This behavior is more realistic than the case of linear fencing cost function. We define the nonlinear function under similar considerations on leakage rate so that a representative function is $G(\gamma) = G_0/(K + \gamma)$, where G_0/K is the cost of perfect fence ($\gamma = 0$), and $G_0 > 0$, $K \ge 0$.

3.4 Numerical Experimentation

In this section, a numerical study is presented to examine the impact of demand leakage rate, γ, and demand variability, σ, on an airline's optimal strategy for fare pricing, seat inventory control, and fencing cost investment. The model-related parameters are adopted from the related numerical study presented in Zhang et al. (2010) in an illustrative example, but these parameters are customized as per the authors' best guess and the additional parameter of airline's cabin capacity, c. Thus, $\alpha_1 = 80$, $\beta_1 = 0.2$, $\alpha_2 = 180$, $\beta_2 = 0.8$, $\mu_1 = \mu_2 = 0$, and $c = 100$. For simplicity, σ_i, $i = \{1,2\}$ are assumed equal for each fare class segment, thus, $\sigma = \sigma_i$ and $\sigma = \{2,5,10,15\}$. In addition to this and consistent with Mostard et al. (2005), the random factor is assumed $\xi_i \in U[-\sqrt{3}\sigma, \sqrt{3}\sigma]$. In a complex problem like the one formulated here, the numerical experimentations are conducted with uniformly distributed price-dependent stochastic demand only (see Zhang et al. 2010). The benefits of fare class creation and differentiated fare pricing are compared with the revenue from the corresponding single fare class, which has a cumulative (equivalent of two fare classes) price-dependent deterministic demand, $260 - p$, and an equivalent single fare class stochastic demand factor $\xi \sim \text{tri}[2\underline{\xi}, 2\overline{\xi}] = \text{tri}[-\sqrt{6}\sigma, \sqrt{6}\sigma]$ with triangular distribution from the convolution of the two uniformly distributed demands (see Zhang et al. 2010). The corresponding single fare class optimal revenues [1]π^* at demand variability $\sigma = \{2,5,10,15\}$ are 15,959.93, 15,745.19, 15,379.58, and 15,010.36, respectively.

A numerical experimentation from the hierarchical optimization approach suggested previously for problem P is presented in Table 3.2. The table reports the optimal decision control parameters, p_1^*, p_2^*, and x_2^* which are prices in each fare class and seat inventory allocation for the discounted fare class segment; notice here that the optimal seat inventory would be simply $x_1^* = c - x_2^*$ and therefore not presented in the table. The airline's revenue from the two fare classes for $\sigma = 2$ is 17,025.81 at no demand leakage (perfect market segmentation), which is about 6.7% superior to the corresponding optimal single segment revenue. Whereas at a higher demand variability and no demand leakage, the revenue gain from two fare classes is noticed about 4.8% superior to the corresponding single fare class revenue. Now, at demand leakage rate of $\gamma = 1$, and a low demand variability, $\sigma = 2$, the

Table 3.2 Numerical Experimentation with Hierarchical Optimization Procedure

σ	γ	x_2^*	p_1^*	p_2^*	$\hat{\pi}^*$
	0	92.79	230.00	142.50	17,025.81
	0.25	86.75	187.32	153.17	16,315.83
2	0.5	98.86	176.97	155.76	16,143.71
	0.75	99.65	172.31	156.92	16,066.16
	1	69.97	169.66	157.59	16,022.04
	0	76.63	230.00	142.50	16,727.04
	0.25	96.46	187.32	153.17	16,072.50
5	0.5	76.26	176.97	155.76	15,913.82
	0.75	81.11	172.31	156.92	15,842.33
	1	99.15	169.66	157.59	15,801.65
	0	87.98	230.00	142.50	16,229.07
	0.25	99.22	187.32	153.17	15,666.94
10	0.5	94.61	176.97	155.76	15,530.67
	0.75	98.66	172.31	156.92	15,469.27
	1	84.55	169.66	157.59	15,434.34
	0	98.45	230.00	142.50	15,731.11
	0.25	99.00	187.32	153.17	15,261.39
15	0.5	98.45	176.97	155.76	15,147.52
	0.75	99.01	172.31	156.92	15,096.21
	1	99.00	169.66	157.59	15,067.02

airline's revenue from two fare classes is noticed only 0.39% superior to the corresponding optimal single fare class revenue. Similarly, at a higher demand variability, $\sigma = 15$, the optimal revenue gains of the airline offering two fare classes are only 0.38% superior to the corresponding optimal single fare class revenue. This clearly leads us to the conclusion that an increase in demand leakage rate, γ, causes a significant effect on the airline's revenue while using market segmentation based on two fare classes compared to a single fare class. The higher demand variability also impacts toward diminishing the revenue gains to an airline, as it can be clearly noticed from the same Table 3.2.

Table 3.3 reports a numerical experimentation with similar findings noticed earlier in sequential optimization approach for problem P. A comparative study of the two methodologies is presented in Figure 3.1, where it can be clearly noticed that both demand leakage rate and demand variability have significant impact on the airline's profitability. In the joint optimization, at $\sigma = 2$, with no demand leakage, the airline

Table 3.3 Numerical Experimentation with Joint Optimization Procedure

σ	γ	x_2^*	p_1^*	p_2^*	$\hat{\pi}^*$
	0	92.26	231.04	144.19	17,068.45
	0.25	74.23	188.17	154.40	16,334.95
2	0.5	68.47	177.79	156.84	16,158.04
	0.75	99.60	173.12	157.93	16,078.49
	1	82.69	170.47	158.54	16,033.27
	0	92.17	232.11	146.46	16,823.91
	0.25	83.35	189.08	155.97	16,114.38
5	0.5	73.20	178.69	158.19	15,944.67
	0.75	80.38	174.02	159.17	15,868.58
	1	72.52	171.37	159.73	15,825.40
	0	97.49	232.73	149.65	16,396.15
	0.25	78.88	189.74	157.98	15,735.01
10	0.5	79.24	179.46	159.85	15,579.25
	0.75	90.22	174.84	160.66	15,509.78
	1	98.79	172.21	161.12	15,470.46
	0	98.94	232.11	152.22	15,950.36
	0.25	86.01	189.59	159.41	15,345.90
15	0.5	86.51	179.50	160.94	15,205.82
	0.75	93.45	174.99	161.60	15,143.68
	1	86.88	172.44	161.97	15,108.61

improves its profitability from 17,025.81 to 17,068.45, which yields about 0.27% revenue increase if the joint optimization procedure is used. However, when demand leakage rate increases to $\gamma = 1$, with a low demand variability of $\sigma = 2$, the optimal revenue of the airline is 16,022.04 in the sequential optimization, while the profitability achieved using the joint optimization is 16,033.27, which is only 0.07% revenue improvement from the sequential framework. At a high demand variability, $\sigma = 15$, and a perfect market segmentation, $\gamma = 0$, the airline's optimal revenue using sequential optimization would be 15,731.11, which is improved with 1.46% to 15,950.36 through the joint optimization approach. Similar to an observation with low demand variability, $\sigma = 2$, when both demand leakage rate and demand variability are higher, $\gamma = 1$ and $\sigma = 15$, the revenue gain from the joint optimization framework compared to sequential optimization reduces to only 0.28%. Thus, we can conclude here that the sequential optimization procedure is quite competitive to the joint optimization procedure.

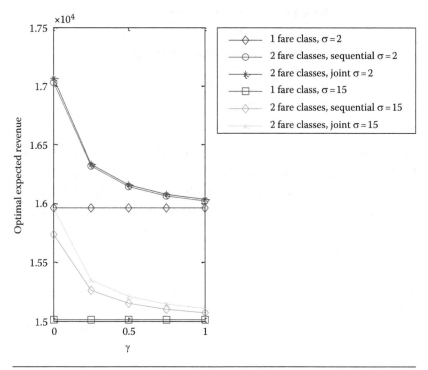

Figure 3.1   Impact of demand leakage and demand variability.

We consider next the extended problem, P', which enables the airline to mitigate or enhance demand leakage rate, γ, between the two fare classes at an additional investment given by $G(\gamma)$. In a study reported in Zhang et al. (2010), we have noticed that the linear fencing, $G(\gamma)$, resulted the firm's optimal decision to either fully control the demand leakage, γ, to zero, or to not invest in fencing. This is due to the fact that the revenue function, π, in problem P' is convex in γ. Alternatively, the nonlinear fencing cost $G(\gamma) = G_0/(K+\gamma)$ is reported in the same study to have a concave revenue function for a firm. Noticeably, when $K = 0$, it is prohibitive for an airline to stop the demand leakage, regardless of its investment, $\lim_{\gamma \to 0} G(\gamma) = \lim_{\gamma \to 0}(G_0/\gamma) \to \infty$. In this study, we have considered the nonlinear fencing cost to optimize the airline's joint decisions on p_1, p_2, x_1, x_2, and γ. The fencing cost function used is given by $G(\gamma) = (100/\gamma)$.

Next, we study the airline's optimal decision of a joint control on p_1, p_2, x_1, x_2, and γ, at various demand variability and with a nonlinear fencing control. In Table 3.4, optimal fencing decision γ is

Table 3.4 Optimal Fencing Decision

σ	p_1^*	p_2^*	x_2^*	γ^*	π^*
2	180.22	156.27	98.67	0.42	15,959.93
5	180.01	157.91	82.60	0.45	15,745.19
10	178.36	160.04	85.15	0.55	15,379.58
15	175.25	161.56	93.37	0.73	15,010.36

determined by a numerical optimal procedure in MATLAB® and Global Optimization Toolbox (The MathWorks, 2013). GlobalSearch procedure from the toolbox with default settings is utilized. It is obvious to notice here that with higher demand variability, an airlines optimal decision on fencing investment would be to keep an increased demand leakage rate.

Figure 3.2a through c illustrate the impact of demand variability, σ, and the optimal fencing decision of the airline. It is obvious to notice here that, with an increase in the demand variability, an airline's optimal investment decision on fencing would be to diminish it

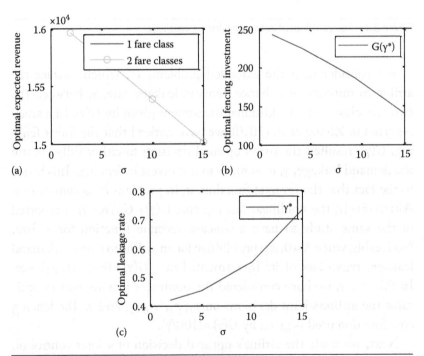

(a) (b) (c)

Figure 3.2 (a)–(c) Impact of optimal fencing decision.

as the demand variability increases. Naturally, it will lead to an airline to increase the optimal demand leakage rate, γ^*.

3.5 Conclusions

In this research, an integrated approach to optimal fare pricing and seat inventory control is presented for an airline that experiences demand leakage. The fences that segment the market demand are considered imperfect. Due to imperfect market segmentation, the airline observes demand leakage from full fare class to the discounted fare class. The research provides models of RM for an airline in the situation when it experiences stochastic price-dependent demands. The models are analyzed to determine an integrated optimal control to fare pricing, seat inventory control and fencing cost decisions.

Numerical experimentations are carried out to underline the impact of both market segmentation and fencing efforts onto the airline's profitability.

The future work directions include investigating the optimal investment strategies in regard to different types of consumer behaviors or specific product features in order to keep the airline immune to demand leakage effects. The present analysis has considered the firm in monopoly only; an interesting avenue, therefore, would be to consider a game theoretic approach to this problem in duopoly or oligopoly.

Appendix 3.A

3.A.1 Derivation of the Revenue Function

$$E[\hat{\pi}] = p_1 \min\{x_1 + x_2 - \min\{x_2, D_2\}, D_1\} + p_2 \min\{x_2, D_2\} \quad (3.22)$$

Notice that $\min\{a,b\} = a - [a-b]^+ = b - [b-a]^+$, where $a, b \in R$ and $[a]^+ = \max\{a,0\}$. Also, $[a-b]^+ = (a-b) - [b-a]^+$ (see Gallego and Moon, 1993; Chen et al., 2004; AlFares and Elmorra, 2005 for details). Furthermore, $E_{\xi_i}(D_i) = z_i = y_i + \mu_i, \forall i = \{1, 2\}$. Thus, we obtain min $\{D_2, x_2\} = x_2 - E_{\xi_2}[x_2 - D_2]^+$ and $\min\{x_1 + x_2 - \min\{x_2, D_2\}, D_1\} = x_1 + E_{\xi_2}[x_2 - D_2]^+ - E_{\xi_1}[x_1 + E_{\xi_2}[x_2 - D_2]^+ - D_1]^+$, and therefore, the revenue from Equation 3.22 becomes:

$$\hat{\pi} = p_1\left(x_1 + E_{\xi_2}\left[x_2 - D_2\right]^+\right) - p_1 E_{\xi_1}\left[x_1 + E_{\xi_2}\left[x_2 - D_2\right]^+ - D_1\right]^+$$

$$- p_2 x_2 - p_2 E_{\xi_2}\left[x_2 - D_2\right]^+ \tag{3.23}$$

Using earlier studies (see Yao, 2002; Yao et al., 2006) in Equation 3.23, we have $E_{\xi_2}[x_2 - D_2]^+ = \int_{\xi_2}^{x_2 - y_2} F_2(\xi_2)d\xi_2.$

And, similarly we can determine the following expression:

$$E_{\xi_1}\left[x_1 + E_{\xi_2}\left[x_2 - D_2\right]^+ - D_1\right]^+\Big|$$

$$= \int_{\xi_1}^{x_1 + \int_{\xi_2}^{x_2 - y_2} F_2(\xi_2)d\xi_2 - y_1}\left(x_1 + \int_{\xi_2}^{x_2 - y_2} F_2(\xi_2)d\xi_2 - y_1 - \xi_1\right)f_1(\xi_1)d\xi_1$$

$$= \int_{\xi_1}^{x_1 + \int_{\xi_2}^{x_2 - y_2} F_2(\xi_2)d\xi_2 - y_1} F_1(\xi_1)d\xi_1$$

Substituting these expressions in Equation 3.23 yields the following revenue function:

$$\hat{\pi} = p_1 x_1 + p_2 x_2 + (p_1 - p_2)\int_{\xi_2}^{x_2 - y_2} F_2(\xi_2)d\xi_2$$

$$- p_1 \int_{\xi_1}^{x_1 + \int_{\xi_2}^{x_2 - y_2} F_2(\xi_2) - y_1} F_1(\xi_1)d\xi_1 \tag{3.24}$$

Proof of Proposition 3.1

1. Applying Karush Kuhn Tucker (KKT) optimality conditions, the Lagrangian function associated to problem DP is:

$$L(p_1, p_2, \lambda) = p_1 z_1 + p_2 z_2 + \lambda(c - z_1 - z_2) \tag{3.25}$$

where
$$z_1 = \alpha_1 - \beta_1 p_1 - \gamma(p_1 - p_2) + \mu_1$$
$$z_2 = \alpha_2 - \beta_2 p_2 + \gamma(p_1 - p_2) + \mu_2$$

The first-order optimality conditions (FOCs) are

$$\frac{\partial L}{\partial p_1} = p_1 \frac{\partial z_1}{\partial p_1} + z_1 + p_2 \frac{\partial z_2}{\partial p_1} - \lambda\left(\frac{\partial z_1}{\partial p_1} + \frac{\partial z_2}{\partial p_1}\right) = 0 \quad (3.26)$$

$$\frac{\partial L}{\partial p_2} = p_1 \frac{\partial z_1}{\partial p_2} + z_2 + p_2 \frac{\partial z_2}{\partial p_2} - \lambda\left(\frac{\partial z_1}{\partial p_2} + \frac{\partial z_2}{\partial p_2}\right) = 0 \quad (3.27)$$

$$\frac{\partial L}{\partial \lambda} = c - z_1 - z_2 \geq 0, \quad \lambda \geq 0, \quad (c - z_1 - z_2)\lambda = 0 \quad (3.28)$$

where
$(\partial z_1/\partial p_1) = -(\beta_1 + \gamma)$
$(\partial z_1/\partial p_2) = (\partial z_2/\partial p_1) = \gamma$
$(\partial z_2/\partial p_2) = -(\beta_2 + \lambda)$

Since $c - z_1 - z_2 = 0$ must be satisfied, therefore, $\lambda > 0$. After the simplification, the KKT optimality conditions become

$$\alpha_1 - 2p_1(\beta_1 + \gamma) + 2p_2\gamma + \lambda\beta_1 + \mu_1 = 0 \quad (3.29)$$

$$\alpha_2 - 2p_2(\beta_2 + \gamma) + 2p_1\gamma + \lambda\beta_2 + \mu_2 = 0 \quad (3.30)$$

$$c - (\alpha_1 + \alpha_2) + \beta_1 p_1 + \beta_2 p_2 - (\mu_1 + \mu_2) = 0 \quad (3.31)$$

2. To prove the joint concavity in p_1 and p_2 of π^d, from *DP*, we explore the Hessian matrix **H**:

$$\mathbf{H} = \begin{bmatrix} \dfrac{\partial^2 \pi^d}{\partial p_1^2} & \dfrac{\partial^2 \pi^d}{\partial p_1 \partial p_2} \\ \dfrac{\partial^2 \pi^d}{\partial p_1 \partial p_2} & \dfrac{\partial^2 \pi^d}{\partial p_2^2} \end{bmatrix} \quad (3.32)$$

Notice here that $(\partial^2 \pi^d/\partial p_1^2) = -2(\beta_1 + \gamma) \leq 0$, $(\partial^2 \pi^d/\partial p_2^2) = -2(\beta_2 + \gamma) \leq 0$ and $(\partial^2 \pi^d/\partial p_1 \partial p_2) = 2\gamma \geq 0$.

Now, **H** is given by

$$\mathbf{H} = \begin{bmatrix} -2(\beta_1 + \gamma) & 2\gamma \\ 2\gamma & -2(\beta_2 + \gamma) \end{bmatrix} \quad (3.33)$$

In order to prove the joint concavity of π^d in p_1 and p_2, the two first principal minors $(\partial^2\pi^d/\partial p_1^2)$ and $(\partial^2\pi^d/\partial p_2^2)$ must be nonpositive, and the second principal minor $|\mathbf{H}| = (\partial^2\pi^d/\partial p_1^2)(\partial^2\pi^d/\partial p_2^2) - (\partial^2\pi^d/\partial p_1\partial p_2)^2$ must be nonnegative. Form Equation 3.33, it can be clearly noticed that both principal minors are negative, and $|\mathbf{H}| = 4(\beta_1\gamma + \beta_2\gamma + \beta_1\beta_2) \geq 0$. This proves the joint concavity of π^d in p_1 and p_2.

Proof of Proposition 3.2

$$1.\ \hat{\pi} = p_1x_1 + p_2x_2 + (p_1 - p_2)\int_{\xi_2}^{x_2-y_2} F_2(\xi_2)d\xi_2$$

$$-p_1 \int_{\xi_1}^{x_1+\int_{\xi_2}^{x_2-y_2} F_2(\xi_2)-y_1} F_1(\xi_1)d\xi_1 \tag{3.34}$$

The revenue function from this equation 3.34 is simplified using the following notations: $I_2 = \int_{\xi_2}^{x_2-y_2} F_2(\xi_2)d\xi_2$ and $I_1 = \int_{\xi_1}^{x_1+\int_{\xi_2}^{x_2-y_2} F_2(\xi_2)-y_1} F_1(\xi_1)d\xi_1$, where $y_1(p_1,p_2,\gamma) = \alpha_1 - \beta_1$ $p_1 - \gamma(p_1 - p_2)$ and $y_2(p_1,p_2,\gamma) = \alpha_2 - \beta_2p_2 + \gamma(p_1 - p_2)$, so that

$$\hat{\pi} = p_1x_1 + p_2x_2 + (p_1 - p_2)I_2 - p_1I_1 \tag{3.35}$$

The FOCs w.r.t. x_i, $i = \{1,2\}$, are

$$\frac{\partial\hat{\pi}}{\partial x_1} = p_1 - p_1\frac{\partial I_1}{\partial x_1} \tag{3.36}$$

$$\frac{\partial\hat{\pi}}{\partial x_2} = p_2 + (p_1 - p_2)\frac{\partial I_2}{\partial x_2} - p_1\frac{\partial I_1}{\partial x_2} \tag{3.37}$$

In these Equations 3.36 and 3.37, $(\partial I_1/\partial x_1) = \Phi_1$, $(\partial I_1/\partial x_2) = \Phi_1\Phi_2$, $(\partial I_2/\partial x_1) = 0$, and $(\partial I_2/\partial x_2) = \Phi_2$, where $\Phi_1 = F_1$ $\left(x_1 + \int_{\xi_2}^{x_2-y_2} F_2(\xi_2) - y_1\right)$, $\Phi_2 = F_2(x_2-y_2)$ and $\phi_1 = f_1\left(x_1 + \int_{\xi_2}^{x_2-y_2}\right.$ $\left. F_2(\xi_2) - y_1\right)$, $\phi_2 = f_2(x_2 - y_2)$. Thus, from Equation 3.36, we have $p_1(1-\Phi_1)=0$, and from Equation 3.37, $p_1\Phi_2(1-\Phi_1)+p_2(1-\Phi_2)=0$. Substituting $p_1(1-\Phi_1)=0$ in Equation 3.37, we obtain $p_2(1-\Phi_2)=0$. Furthermore, it is obvious to notice that $p_2 > 0$, which yields the optimality condition such that $\Phi_2=1$. This translates into $F_2(x_2-y_2)=1$ and since $\bar{\xi}_i = \sqrt{3}\sigma$, which will result, $F_2^{-1}(1) = \sqrt{3}\sigma$, and thus optimal seat allocation for discounted fare class would be $x_2^* = y_2 + \sqrt{3}\sigma$.

2. The joint concavity of $\hat{\pi}$ in x_1 and x_2 is satisfied if the Hessian matrix $\mathbf{H}$ is negative semidefinite:

$$\mathbf{H} = \begin{bmatrix} \dfrac{\partial^2\hat{\pi}}{\partial x_1^2} & \dfrac{\partial^2\hat{\pi}}{\partial x_1 \partial x_2} \\[2ex] \dfrac{\partial^2\hat{\pi}}{\partial x_1 \partial x_2} & \dfrac{\partial^2\hat{\pi}}{\partial x_2^2} \end{bmatrix} \qquad (3.38)$$

The first principal minor conditions for the joint concavity of $\hat{\pi}$ are $(\partial^2\hat{\pi}/\partial x_1^2) = -p_1\phi_1 \leq 0$ and $(\partial^2\hat{\pi}/\partial x_2^2) = -\phi_2(p_1\Phi_1 - (p_1 - p_2)) - p_1\phi_1\Phi_2^2 \leq 0$, given $p_1\Phi_1-(p_1-p_2) \geq 0$. Next, the second principal minor is $|\mathbf{H}| = (\partial^2\hat{\pi}/\partial x_1^2)(\partial^2\hat{\pi}/\partial x_2^2) - (\partial^2\hat{\pi}/\partial x_1 \partial x_2)^2$, where $(\partial^2\hat{\pi}/\partial x_1 \partial x_2) = -p_1\phi_1\Phi_2$, and therefore, $|\mathbf{H}| = p_1\phi_1\phi_2(p_1\Phi_1-(p_1-p_2))$. To prove the joint concavity of $\hat{\pi}$ w.r.t. capacity allocations x_i, $\forall i = \{1,2\}$, the condition for second principal minor $p_1\phi_1\phi_2(p_1\Phi_1-(p_1-p_2)) \geq 0$ must be satisfied, which implies a similar condition established for the principal minors of $\mathbf{H}$, that is, $p_1\Phi_1-(p_1-p_2) \geq 0$.

Proof of Proposition 3.3

1. Joint concavity of π w.r.t. p_1 and p_2 is satisfied if $\mathbf{H}$ is negative semidefinite, where

$$\mathbf{H} = \begin{bmatrix} \dfrac{\partial^2\pi}{\partial p_1^2} & \dfrac{\partial^2\pi}{\partial p_1 \partial p_2} \\[2ex] \dfrac{\partial^2\pi}{\partial p_1 \partial p_2} & \dfrac{\partial^2\pi}{\partial p_2^2} \end{bmatrix} \qquad (3.39)$$

The first-order derivatives of π from Equation 3.8 w.r.t. p_1 and p_2 are

$$\frac{\partial\pi}{\partial p_1} = x_1 + I_2 + \left(p_1 - p_2\right)\frac{\partial I_2}{\partial p_1} - I_1 - p_1\frac{\partial I_1}{\partial p_1} = 0 \qquad (3.40)$$

$$\frac{\partial \pi}{\partial p_2} = x_2 - I_2 + \left(p_1 - p_2 \right) \frac{\partial I_2}{\partial p_2} - p_1 \frac{\partial I_1}{\partial p_2} = 0 \qquad (3.41)$$

Using the previous relations,

$$\frac{\partial y_1}{\partial p_1} = -(\beta_1 + \gamma), \quad \frac{\partial y_1}{\partial p_2} = \frac{\partial y_2}{\partial p_1} = \gamma, \quad \text{and} \quad \frac{\partial y_2}{\partial p_2} = -(\beta_2 + \gamma)$$

and the derivatives

$$\frac{\partial I_1}{\partial p_1} = \Phi_1 \left(\beta_1 + \gamma (1 - \Phi_2) \right), \quad \frac{\partial I_1}{\partial p_2} = \Phi_1 \left(\beta_2 \Phi_2 - \gamma (1 - \Phi_2) \right)$$

$$\frac{\partial I_2}{\partial p_2} = \Phi_2 \left(\beta_2 + \gamma \right), \quad \frac{\partial I_2}{\partial p_1} = -\Phi_2 \gamma$$

we can write Equations 3.40 and 3.41 as

$$\frac{\partial \pi}{\partial p_1} = x_1 - \left(p_1 - p_2 \right) \Phi_2 \gamma - p_1 \Phi_1 \left(\beta_1 + \gamma (1 - \Phi_2) \right) - I_1 + I_2 = 0$$

$$(3.42)$$

$$\frac{\partial \pi}{\partial p_2} = x_2 - I_2 + \left(p_1 - p_2 \right) \Phi_2 \left(\beta_2 + \gamma \right)$$

$$- p_1 \Phi_1 \left(\beta_2 \Phi_2 - \gamma (1 - \Phi_2) \right) = 0 \qquad (3.43)$$

Hessian's first principal minors are given by

$$\frac{\partial^2 \pi}{\partial p_1^2} = 2 \frac{\partial I_2}{\partial p_1} + \left(p_1 - p_2 \right) \frac{\partial^2 I_2}{\partial p_1^2} - 2 \frac{\partial I_1}{\partial p_1} - p_1 \frac{\partial^2 I_1}{\partial p_1^2} \qquad (3.44)$$

$$\frac{\partial^2 \pi}{\partial p_2^2} = -2 \frac{\partial I_2}{\partial p_2} + \left(p_1 - p_2 \right) \frac{\partial^2 I_2}{\partial p_2^2} - p_1 \frac{\partial^2 I_1}{\partial p_2^2} \qquad (3.45)$$

And, the partial derivative is

$$\frac{\partial^2 \pi}{\partial p_1 \partial p_2} = \frac{\partial I_2}{\partial p_2} - \frac{\partial I_2}{\partial p_1} + \left(p_1 - p_2 \right) \frac{\partial^2 I_2}{\partial p_1 \partial p_2} - \frac{\partial I_1}{\partial p_2} - p_1 \frac{\partial^2 I_1}{\partial p_1 \partial p_2}$$

$$(3.46)$$

where the partial second-order derivatives of I_i, $\forall i = \{1,2\}$, w.r.t. p_i, $\forall i = \{1,2\}$ are:

$$\frac{\partial^2 I_1}{\partial p_1^2} = \Phi_1 \phi_2 \gamma^2 + \phi_1 \left(\beta_1 + \gamma \left(1 - \Phi_2 \right) \right)^2$$

$$\frac{\partial^2 I_1}{\partial p_2^2} = \Phi_1 \phi_2 \left(\beta_2 + \gamma \right)^2 + \phi_1 \left(\Phi_2 \beta_2 - \gamma \left(1 - \Phi_2 \right) \right)^2$$

$$\frac{\partial^2 I_1}{\partial p_1 \partial p_2} = \phi_1 \left(\beta_1 + \gamma \left(1 - \Phi_2 \right) \right)\left(\beta_2 \Phi_2 - \gamma \left(1 - \Phi_2 \right) \right)$$

$$- \Phi_1 \phi_2 \gamma \left(\beta_2 + \gamma \right)$$

$$\frac{\partial^2 I_2}{\partial p_2^2} = \phi_2 \left(\beta_2 + \gamma \right)^2, \quad \frac{\partial^2 I_2}{\partial p_1^2} = \phi_2 \gamma^2, \quad \frac{\partial^2 I_2}{\partial p_1 \partial p_2} = -\phi_2 \gamma \left(\beta_2 + \gamma \right)$$

For further simplification, we use the following notations: $t_1 = \beta_1 + \gamma(1 - \Phi_2)$, $t_2 = \Phi_2 \beta_2 - \gamma(1 - \Phi_2)$, and $t_3 = p_1 \Phi_1 - (p_1 - p_2)$. It is obvious to notice that $t_1 \geq 0$, and with the findings from Proposition 3.2, we find that $t_3 \geq 0$. To further simplify, we assume that $t_2 \geq 0$. This yields t_1, t_2, and $t_3 \geq 0$. Thus, Equations 3.44 through 3.46 can be reduced using t_1, t_2, and t_3 notations to the following expressions:

$$\frac{\partial^2 \pi}{\partial p_1^2} = -2\Phi_2 \gamma + \left(p_1 - p_2 \right) \phi_2 \gamma^2 - 2\Phi_1 \left(\beta_1 + \gamma \left(1 - \Phi_2 \right) \right)$$

$$- p_1 \left(\Phi_1 \phi_2 \gamma^2 + \phi_1 \left(\beta_1 + \gamma \left(1 - \Phi_2 \right) \right)^2 \right)$$

$$= -2\Phi_2 \gamma - 2\Phi_1 t_1 - p_1 \phi_1 t_1^2 - \phi_2 \gamma^2 t_3 \qquad (3.47)$$

$$\frac{\partial^2 \pi}{\partial p_2^2} = -2\Phi_2 \left(\beta_2 + \gamma \right) + \left(p_1 - p_2 \right) \phi_2 \left(\beta_2 + \gamma \right)^2$$

$$- p_1 \left(\Phi_1 \phi_2 \left(\beta_2 + \gamma \right)^2 + \phi_1 \left(\Phi_2 \beta_2 - \gamma \left(1 - \Phi_2 \right) \right)^2 \right)$$

$$= -2\Phi_2 \left(\beta_2 + \gamma \right) - p_1 \phi_1 t_2^2 - \phi_2 \left(\beta_2 + \gamma \right)^2 t_3 \qquad (3.48)$$

$$\frac{\partial^2 \pi}{\partial p_1 \partial p_2} = \Phi_2\left(\beta_2 + \gamma\right) + \Phi_2\gamma - \phi_2\gamma\left(\beta_2 + \gamma\right)\left(p_1 - p_2\right)$$

$$-\Phi_1\left(\beta_2\Phi_2 - \gamma\left(1 - \Phi_2\right)\right)$$

$$-p_1\left(\phi_1\left(\beta_1 + \gamma\left(1 - \Phi_2\right)\right)\left(\beta_2\Phi_2 - \gamma\left(1 - \Phi_2\right)\right)\right.$$

$$\left. -\Phi_1\phi_2\gamma\left(\beta_2 + \gamma\right)\right)$$

$$= \Phi_2\left(\beta_2 + 2\gamma\right) - \Phi_1 t_2 - p_1\phi_1 t_1 t_2 + \phi_2\gamma\left(\beta_2 + \gamma\right)t_3$$

$$(3.49)$$

It is clear to notice from Equations 3.47 and 3.48 that the first principal minors are both nonpositive. Now, we need to show the second principal minor sign is positive; therefore, we need to prove

$$|\mathbf{H}| = \frac{\partial^2 \pi}{\partial p_1^2}\frac{\partial^2 \pi}{\partial p_2^2} - \left(\frac{\partial^2 \pi}{\partial p_1 \partial p_2}\right)^2 \ge 0$$

where $|\mathbf{H}|$ is determined after some simplification as

$$|\mathbf{H}| = \left(p_1\phi_1 t_1^2 + \phi_2\gamma^2 t_3 + 2\left(\Phi_2\gamma + \Phi_1 t_1\right)\right)$$

$$\times \left(p_1\phi_1 t_2^2 + \phi_2(\beta_2 + \gamma)^2 t_3 + 2\Phi_2\left(\beta_2 + \gamma\right)\right)$$

$$-\left(\Phi_2\left(\beta_2 + 2\gamma\right) + \phi_2\gamma\left(\beta_2 + \gamma\right)t_3 - \left(\Phi_1 t_2 + p_1\phi_1 t_1 t_2\right)\right)^2$$

$$(3.50)$$

Given that $p_1, \Phi_1, \phi_1, p_2, \Phi_2, \phi_2, t_1, t_2, t_3 \ge 0$, we can achieve a lower bound on $|\mathbf{H}|$ established in Equation 3.50 by ignoring some positive terms. While simplifying the rest of the terms, we obtain the following reduced form:

$$|\mathbf{H}| \ge \left(p_1\phi_1 t_1^2\right)\left(p_1\phi_1 t_2^2\right) - \left(\beta_2 + 2\gamma + \gamma(\beta_2 + \gamma)t_3\right)^2 \quad (3.51)$$

Therefore, the condition for joint concavity will be

$$\left(p_1\phi_1 t_1^2\right)\left(p_1\phi_1 t_2^2\right) - \left(\beta_2 + 2\gamma + \gamma(\beta_2 + \gamma)t_3\right)^2 \ge 0 \quad (3.52)$$

which can be further written as

$$\left(p_1 \phi_1 t_1 t_2 + \left(\beta_2 + 2\gamma + \gamma \left(\beta_2 + \gamma \right) t_3 \right) \right)$$

$$\times \left(p_1 \phi_1 t_1 t_2 - \left(\beta_2 + 2\gamma + \gamma \left(\beta_2 + \gamma \right) t_3 \right) \right) \geq 0 \qquad (3.53)$$

Finally, the necessary condition for joint concavity of π will be $p_1 \phi_1 t_1 t_2 - (\beta_2 + 2\gamma + \gamma(\beta_2 + \gamma)t_3) \geq 0$. There can be other possibilities that may also guarantee the joint concavity of π; however, this chapter only focuses on the single possibility presented in this proof.

2. The Lagrangian function of nonlinear problem P' is

$$L\left(x_1, x_2, p_1, p_2, \gamma, \lambda \right) = p_1 x_1 + p_2 x_2 + \left(p_1 - p_2 \right) I_2 - p_1 I_1 - G(\gamma)$$

$$+ \lambda \left(c - x_1 - x_2 \right)$$

The KKT optimality conditions are

$$\frac{\partial L}{\partial x_1} = p_1 \left(1 - \Phi_1 \right) - \lambda = 0 \qquad (3.54)$$

$$\frac{\partial L}{\partial x_2} = p_2 + \left(p_1 - p_2 \right) \Phi_2 - p_1 \Phi_1 \Phi_2 - \lambda = 0 \qquad (3.55)$$

$$\frac{\partial L}{\partial p_1} = x_1 - I_1 + I_2 - \left(p_1 - p_2 \right) \Phi_2 \gamma$$

$$- p_1 \Phi_1 \left(\beta_1 + \gamma \left(1 - \Phi_2 \right) \right) = 0 \qquad (3.56)$$

$$\frac{\partial L}{\partial p_2} = x_2 - I_2 + \left(p_1 - p_2 \right) \Phi_2 \left(\beta_2 + \gamma \right)$$

$$- p_1 \Phi_1 \left(\beta_2 \Phi_2 - \gamma \left(1 - \Phi_2 \right) \right) = 0 \qquad (3.57)$$

$$\frac{\partial L}{\partial \gamma} = -\Phi_2 \left(p_1 - p_2 \right)^2 - p_1 \Phi_1 \left(p_1 - p_2 \right) \left(1 - \Phi_2 \right)$$

$$- \frac{\partial G(\gamma)}{\partial \gamma} = 0 \qquad (3.58)$$

$$\frac{\partial L}{\partial \lambda} = c - x_1 - x_2 = 0 \qquad (3.59)$$

Recalling for Equations 3.54 through 3.58, the notations are

$$\Phi_1 = F_1\left(x_1 + \int_{\underline{\xi}_2}^{x_2 - y_2} F_2(\xi_2) - y_1\right), \quad \Phi_2 = F_2(x_2 - y_2)$$

$$I_2 = \int_{\underline{\xi}_2}^{x_2 - y_2} F_2(\xi_2)\,d\xi_2, \quad I_1 = \int_{\underline{\xi}_1}^{x_1 + \int_{\underline{\xi}_2}^{x_2 - y_2} F_2(\xi_2) - y_1} F_1(\xi_1)\,d\xi$$

Therefore, to determine the optimal solution $(x_1^*, x_2^*, p_1^*, p_2^*, \gamma^*)$, we will have to solve the following system of nonlinear equations:

$$p_1(1 - \Phi_1) - p_2 + \Phi_2\big(p_1\Phi_1 - (p_1 - p_2)\big) = 0 \qquad (3.60)$$

$$x_1 - I_1 + I_2 - (p_1 - p_2)\Phi_2\gamma - p_1\Phi_1\big(\beta_1 + \gamma(1 - \Phi_2)\big) = 0 \quad (3.61)$$

$$x_2 - I_2 + (p_1 - p_2)\Phi_2(\beta_2 + \gamma) - p_1\Phi_1\big(\beta_2\Phi_2 - \gamma(1 - \Phi_2)\big) = 0$$
$$\qquad (3.62)$$

$$-\Phi_2(p_1 - p_2)^2 - p_1\Phi_1(p_1 - p_2)(1 - \Phi_2) - \frac{\partial G}{\partial \gamma} = 0 \quad (3.63)$$

$$c - x_1 - x_2 = 0 \qquad (3.64)$$

Proof of Proposition 3.4

We consider the linear fencing cost function $G(\gamma) = G_0 - (G_0/K)\gamma$, where $G_0 > 0$, $K > 0$, and G_0 is the cost of null leakage when perfect fences are achieved so that $\gamma = 0$. When there is no initiative to invest in fencing, $G(\gamma) = 0$. The rate of change in $G(\gamma)$ w.r.t. γ is $(\partial G/\partial \gamma) = -(G_0/K)$ and $(\partial^2 G/\partial \gamma^2) = 0$ due to linear $G(\gamma)$. Notice here that $G(\gamma = 0) = G_0$ and $G(\gamma = K) = 0$.

1. Recalling the revenue function, π, from Equation 3.8

$$\pi(x_i, p_i, \gamma) = p_1 x_1 + p_2 x_2 + (p_1 - p_2) I_2 - p_1 I_1 - G(\gamma)$$

The partial derivatives of π w.r.t. γ are

$$\frac{\partial \pi}{\partial \gamma} = (p_1 - p_2) \frac{\partial I_2}{\partial \gamma} - p_1 \frac{\partial I_1}{\partial \gamma} - \frac{\partial G(\gamma)}{\partial \gamma}$$

$$= -\Phi_2 (p_1 - p_2)^2 - p_1 \Phi_1 (p_1 - p_2)(1 - \Phi_2) - \frac{G_0}{K} \quad (3.65)$$

$$\frac{\partial^2 \pi}{\partial \gamma^2} = (p_1 - p_2) \cdot \frac{\partial^2 I_2}{\partial \gamma^2} - p_1 \cdot \frac{\partial^2 I_1}{\partial \gamma^2}$$

$$= \phi_2 (p_1 - p_2)^3 - p_1 (p_1 - p_2)^2 \left(\phi_1 (1 - \Phi_2)^2 + \Phi_1 (1 - \phi_2) \right)$$

$$= (p_1 - p_2)^2 \left(\phi_2 (p_1 - p_2) - p_1 (\phi_1 (1 - \Phi_2)^2 + \Phi_1 (1 - \phi_2)) \right)$$
$$(3.66)$$

where

$$\frac{\partial I_2}{\partial \gamma} = -\Phi_2 (p_1 - p_2), \quad \frac{\partial I_1}{\partial \gamma} = \Phi_1 (p_1 - p_2)(1 - \Phi_2)$$

$$\frac{\partial^2 I_2}{\partial \gamma^2} = \phi_2 (p_1 - p_2)^2, \quad \frac{\partial^2 I_1}{\partial \gamma} = (p_1 - p_2)^2 \left(\phi_1 (1 - \Phi_2)^2 \right.$$

$$\left. + \Phi_1 (1 - \phi_2) \right)$$

$$\Phi_1 = F_1 \left(x_1 + \int_{\xi_2}^{x_2 - y_2} F_2(\xi_2) - y_1 \right), \quad \Phi_2 = F_2 (x_2 - y_2)$$

$$\phi_1 = f_1 \left(x_1 + \int_{\xi_2}^{x_2 - y_2} F_2(\xi_2) - y_1 \right), \quad \phi_2 = f_2 (x_2 - y_2)$$

From Equation 3.65, we can determine γ^* by solving $(p_1 - p_2)(\Phi_2 (p_1 - p_2) + p_1 \Phi_1 (1 - \Phi_2)) + (G_0/K) = 0$, given that p_i, x_i,

$\forall i = \{1,2\}$ are known. Notice from Equation 3.65 that the total expected revenue, π, is nonincreasing in leakage rate, γ, as $(\partial\pi/\partial\gamma) \leq 0$ for $0 \leq \gamma \leq K$. From Equation 3.66, π is quasiconcave in γ if $\phi_2(p_1 - p_2) - p_1(\phi_1(1 - \Phi_2)^2 + \Phi_1(1 - \phi_2)) \leq 0$.

Acknowledgment

This publication was made possible by NPRP grant # 5-023-05-006 from the Qatar National Research Fund (a member of Qatar Foundation). The statements made herein are solely the responsibility of the authors.

References

AlFares, H. and Elmorra, H. (2005). The distribution-free newsboy problem: Extensions to the shortage penalty case, *International Journal of Production Economics* 93/94, 465–477.

Anon. (n.d.). The theory and practice of revenue management [online]. Available at: http://www.springer.com/business+&+management/operations+research/book/978-1-4020-7701-2 (accessed January 26, 2014).

Bell, P.C. (1998). Revenue management: That's the ticket, *OR/MS Today*, 25(2).

Chen, F.Y., Yan, H., and Yao, Y. (2004). A newsvendor pricing game, *IEEE Transactions on Systems, Man and Cybernetics, Part A: Systems and Humans* 34(4), 450–456.

Chiang, W.C., Chen, J.C.H., and Xu, X. (2007). An overview of research on revenue management: Current issues and future research, *International Journal of Revenue Management* 1(1), 97–128.

Chiang, W.K. and Monahan, G.E. (2005). Managing inventories in a two-echelon dualchannel supply chain, *European Journal of Operational Research* 162(2), 325–341.

Choi, S.C. (1996). Pricing competition in a duopoly common retailer channel, *Journal of Retailing* 72(2), 117–134.

Cote, J.P., Marcotte, P., and Savard, G. (2003). A bilevel modelling approach to pricing and fare optimisation in the airline industry, *Journal of Revenue Management and Pricing* 2, 23–36.

Feng, Y. and Xiao, B. (2001). A dynamic airline seat inventory control model and its optimal policy, *Operations Research* 49, 939–949.

Gallego, G. and Moon, I. (1993). The distribution free newsboy problem: Review and extensions, *Journal of Operational Research Society* 44, 825–834.

Hanks, R., Cross, R., and Noland, P. (2002). Discounting in the hotel industry, *Cornell Hotel and Restaurant Administration Quarterly* 43, 94–103.

Kimes, S.E. (2002). Perceived fairness of yield management, *Cornell Hotel and Restaurant Administration Quarterly* 43, 21–30.

Li, M.Z.F. (2001). Pricing non-storable perishable goods by using a purchase restriction with an application to airline fare pricing, *European Journal of Operational Research* 134(3), 631–647.

Littlewood, K. (1972). Forecasting and control of passenger booking. *AGIFORS 12th Annual Symposium Proceedings*, Nathanya, Israel.

McGill, J.I. and Van Ryzin, G.J. (1999). Revenue management: Research overview and prospects, *Transportation Science* 33, 233–256.

Mostard, J., Koster, R., and Teunter, R. (2005). The distribution-free newsboy problem with resalable returns, *International Journal of Production Economics* 97, 329–342.

Petruzzi, N.C. and Dada, M. (1999). Pricing and the news vendor problem: A review with extensions, *Operations Research* 47, 183–194.

Philips, R.L. (2005). *Pricing and Revenue Optimization* Stanford, CA: Stanford University Press.

Raza, S.A. and Akgunduz, A. (2008). An airline revenue management fare pricing game with seats allocation, *International Journal of Revenue Management* 2(1), 42–62.

Raza, S.A. and Akgunduz, A. (2010). The impact of fare pricing cooperation in airline revenue management, *International Journal of Operational Research* 7(3), 277–296.

Smith, N.R., Martinez-Flores, J.L., and Cardenas-Barron, L.E. (2007). Analysis of the benefits of joint price and order quantity optimisation using a deterministic profit maximization model, *Production Planning and Control* 18(4), 310–318.

MATLAB and Global Optimization Toolbox R. (2013a). The MathWorks, Inc., Natick Massachusetts, United States, Software available at http://www.mathworks.com/products/matlab/

Weatherford, L.R. (1997). Using prices more realistically as decision variables in perishable-asset revenue management problems, *Journal of Combinatorial Optimization* 1, 277–304.

Yao, L. (2002). *Supply Chain Modeling: Pricing, Contracts and Coordination.* The Chinese University of Hong Kong, Shatin, Hong Kong.

Yao, L., Chen, Y.F., and Yan, H. (2006). The newsvendor problem with pricing: Extension, *International Journal of Management Science and Engineering Management* 1(1), 3–16.

Zhang, M. and Bell, P. (2010). Price fencing in the practice of revenue management: An overview and taxonomy, *Journal of Revenue and Pricing Management* 11(2), 146–159.

Zhang, M. and Bell, P.C. (2007). The effect of market segmentation with demand leakage between market segments on a firm's price and inventory decisions, *European Journal of Operational Research* 182(2), 738–754.

Zhang, M., Bell, P.C., Cai, G., and Chen, X. (2010). Optimal fences and joint price and inventory decisions in distinct markets with demand leakage, *European Journal of Operational Research* 204, 589–596.

4

BI-OBJECTIVE BERTH–CRANE ALLOCATION PROBLEM IN CONTAINER TERMINALS

DENIZ OZDEMIR AND EVRIM URSAVAS

Contents

4.1 Motivation

Transportation via sea continues to rise as a result of the increasing demand due to its advantages over other transportation modes in terms of cost and security. Actually, as of 2013, seaborne trade accounted for 80% of global trade in terms of volume (UNCTAD, 2013), and since 2006, it counts for 70.1% in terms of value (Rodrigue et al., 2009). Due to this trend toward sea transportation, efficient port management has become a major issue for port owners and shipping companies. Typical operations in a port consist of allocation of berths to arriving vessels, allocation of cranes to docked vessels at the quayside, routing of internal transportation vehicles, storage space assignment, and gantry crane deployment at the yard side. Berth allocation problem (BAP) consists of assigning berth spaces to the incoming vessels. Crane allocation problem (CAP) is the determination of the assignment sequence of cranes to a container ship. Both problems on the quayside have received significant attention from researchers (Bierwirth and Meisel, 2010). More often, these two problems are studied separately in the literature, resulting in suboptimal solutions. To find more realistic solutions, researchers offer solutions that combine the two problems.

Port operations involve multiple parties such as ship owners, crane operators, port management, and government officers. By its nature, each party has its own concerns and requirements that need to be addressed in a decision-making process. Hence, the berth allocation and crane scheduling problem requires that the decision makers consider multiple objectives at a time, which, again, adds to the complexity of the problem. An essential concern to deliberate is the fact that objectives such as minimizing vessel service time and maximizing crane utilizations frequently conflict with each other. That is, the decision maker is forced to attain a balance among those conflicting objectives. However, recent literature on the berth and crane scheduling problem does not provide adequate support to resolve the issue.

With those in mind, this study attempts to simultaneously determine the berthing and crane allocations under multiple objectives. In principle, with the existence of more than one objective, we would expect to have a set of optimal solutions instead of a single optimal solution. Therefore, our approach will be to determine these set of solutions, also referred as Pareto optimal solutions, in order to determine Pareto efficient frontier. Following this multisolution approach offers the decision maker the flexibility of adjusting the balance within conflicting objectives.

We may depict the contributions of this chapter as twofold. First, we extend the existing literature by embracing more practical assumptions to better represent the real-world implementation. Second, we formulate a bi-objective integer problem and propose an ε-constraint method-based solution algorithm to acquire the nondominated berth–crane assignments and schedules as Pareto optimal front.

The structure of the remaining part of the chapter is as follows: the following section is dedicated to the related studies in the literature. Section 4.3 is devoted to the mathematical model description of the problem. Section 4.4 puts forward our solution methodology based on ε-constraint method. Section 4.5 reports the computational experiments via a case study. Finally, Section 4.6 concludes the study and states future research directions.

4.2 Related Work

BAPs and CAPs aim to display the berthing position and service sequence of all the vessels; hence, it denotes an assignment and scheduling problem structure. In most of the studies in literature, crane allocation is planned after berthing the ship, which results in suboptimal solutions. Our focus in this review process will put an emphasis on studies that simultaneously tackle both problems.

Work by Zhou and Kang (2008) has used the genetic algorithm to search through the solution space and compared it with the greedy algorithm for the BAP and CAP with stochastic arrival and handling times. The genetic algorithm proposed has significantly improved the greedy algorithm solutions so as to minimize the average waiting time of containerships in terminal. Zhang et al. (2010) use the subgradient optimizations technique to solve the problem with the aim of minimizing the weighted sum of the handling costs of containers. Review work provided by Bierwirth and Meisel (2010) as well as Carlo et al. (2014) presents state-of-the-art research on the topic that jointly tackles berth allocation and crane scheduling.

Recent studies that maintain a multiobjective approach can be summarized as follows: Imai et al. (2007) address the problems with a bi-objective approach that considers the minimization of delay of ships' departure and minimization of the total service time. They use the weighting method that combines all objectives into a single one by assigning weights and by changing the weights in a systematic fashion. They so form the noninferior solution set. Golias et al. (2009) use the multiobjective approach to differentiate the service level given to customers with different priorities. Total service time minimization is realized separately for different levels of customer preferences. Their solution approach is by the use of evolutionary algorithms. In their latter work, they propose a nonnumerical ranking preference method to select the efficient berth schedule (Golias et al., 2010). Cheong et al. (2010a) model the BAP so as to minimize the three objectives of makespan, waiting time, and degree of deviation from a predetermined priority schedule. They use a multiobjective evolutionary algorithm to find the Pareto efficient frontier. However, studies discussed here do not tackle the CAP, and the solution set they provide is not guaranteed to be optimal.

Cheong et al. (2010b) extend the literature by incorporating the crane scheduling problem. They design their problem to solve the two objectives of waiting time and handling time of ships. They as well use the multiobjective evolutionary algorithm approach to model the port conditions at the Pasir Panjang container terminal.

The most related work to our study belongs to Liang et al. (2011). In their bi-objective crane and berth allocation model, they propose a hybrid genetic algorithm to minimize the sum of the handling time of containers and the number of crane movements concurrently. Their computational experiments are realized by a real-world case study of Shanghai container terminal.

In this chapter, we approach the berth–crane scheduling problem concurrently, while considering two objectives of total service time minimization and crane setup minimization. Our crane-related objective differs from the work of Liang et al. (2011), in that their approach aims to avoid the probable crane splits among berths. However, there is no cost incurred for a vessel to be served by crane j at time t, then crane j' at time $t + 1$ and crane j again at time $t + 2$ as long as the cranes are at the same berth. We, in turn, by minimizing the crane setups for each vessel, incorporate the potential cost of crane splitting together with their setup cost, giving a more detailed analysis of crane activities. Moreover, we lead the former work in the perspective of real-world representation. In our model, cranes differ in terms of their technical specifications regarding their container handling rates. Hence, particular cranes may be favored to another in convenient cases. Berth length restrictions and vessel length compatibility issues are also reflected in our model. To the best of our knowledge, this is the first attempt to provide the optimum Pareto efficient frontier for the considered problem.

As to the exact methods for the solution of multiobjective combinatorial optimization problems, several scalarization techniques may be used. The most popular is by the use of the weighted sum approach where different objectives are aggregated through weighted sums. Although the efficient solutions found by the technique may be valid for linear programming problems, due to the discrete structure of the combinatorial problems, the results may not compromise the whole efficient solution set for the considered problem. The consideration of these nonsupported efficient solutions, which are not optimal for any weighted sum of the objectives, becomes crucial when there are more than one sum objective in contrast to cases where at most one sum objective is present and the others are bottleneck objectives. Another approach followed is the compromise solution method, where the distance to a reference point is minimized. The reference point is defined by the separate minima of each objective. Obviously, for conflicting objectives, it is not possible to obtain the minimum limits simultaneously. For bi-objective problems, the use of ranking methods is popular. As required by the technique, the computation of nadir point is difficult to obtain when there are more than two objectives. For the comprehensive description of the available methods, readers may refer to Ehrgott and Gandibleux (2002). For the case of two objectives, the two-phase method is described as a general framework. In two-phase method, the supported efficient solutions are found by the use of scalarization methods in the first phase, and then the nonsupported efficient solutions are found by problem-specific techniques in the second phase.

The solution approach we use to solve our bi-objective integer problem is an iterative algorithm incorporating the branch-and-cut solution embedded in ε-constraint method. ε-Constraint method is one of the well-known techniques to solve multiobjective optimization problems. In ε-constraint method, instead of combining the objectives with weights, only one of the original objectives is minimized while the others are rearranged as constraints. An extensive discussion of the method can be found in Ehrgott (2005).

4.3 Model Description

This study attempts to simultaneously determine the berthing and crane allocations under two objectives. The wharf is modeled to be discrete, that is, it represents a collection of partitioned sections. Different types of cranes with different handling rates are considered. Handling time and the number of cranes to be assigned to the ship are not known in advance. Handling time depends on the type and the number of cranes allocated to a vessel, which is dynamic throughout the service time. For instance, a vessel can start to be served by only one crane and end up being served by three cranes. Therefore, the ships do not have to wait until a specified number of cranes are available. This prevents suboptimal solutions resulting from misleading crane unavailability assumption.

We now present the bi-objective optimization model for solving simultaneous berth–vessel–crane allocation problem. The basic assumptions of the model can be summarized as follows.

4.3.1 Assumptions

1. There are discrete berths with specified lengths. A vessel may be assigned to any of the available berths as long as the vessel length fits to the berth length.
2. There are cranes with different technology that give service with varying handling rates.
3. Some of the cranes are mobile, in a sense that cranes can be assigned to any berth and any vessel in any order.
4. Crane allocation is dynamic throughout the handling period of a vessel. The number and the type of cranes assigned are flexible, and vessel handling time is dependent on crane allocations.
5. A vessel cannot be given service before its arrival.
6. Each different crane allocation incurs a cost.
7. There are a maximum allowable number of cranes that can be assigned to a vessel.

The indices, parameters, decision variables, and the integer linear programming model are defined as follows.

4.3.2 Notation

Indices

$i = (1, ..., I)$ set of vessels

$j = (1, ..., J)$ set of cranes, where first p cranes are static and last $J-p$ cranes are assumed to be portable

$k = (1, ..., K)$ set of berths

$t = (1, ..., T)$ time periods

Input Parameters

l_i: Vessel length including the safety margin for the vessel
Q_k: Length of berth k
a_i: Arrival time of vessel i
N_{i0}: Number of containers initially on the vessel
U: Maximum number of cranes that can be assigned to a vessel simultaneously
R_j: Container handling rate of jth crane

For modeling purposes, we define two constants:

M: Large constants
m: Constant $0 \le m \le 1$

Decision Variables

y_{ijtk}: 1 if crane j is allocated to vessel i at time t at berth k and 0 otherwise
BV_{itk}: 1 if vessel i is assigned to berth k at time t
N_{it}: Total number of containers on vessel i at time t
Δ_{ik}: 1 if vessel i is assigned to berth k
YH_{it}: 1 if vessel i is served at time t
PH_{it}: 1 if vessel i has remaining containers at time t
CR_{ijt}: 1 if crane j will start serving vessel i at time $t + 1$
$TempH_{it}$: Auxiliary variable that realizes the logical connection between y_{ijtk} and YH_{it}

4.3.3 Model

$$f_1(\textit{time}): \ \min \sum_i \sum_{t=a_i} PH_{it}$$

$$f_2(\textit{setup}): \ \min \sum_i \sum_j \sum_{t=a_i} CR_{ijt}$$

$$l_i \cdot y_{ijtk} \le Q_k \quad \forall i,j,t,k \qquad (4.1)$$

$$\sum_k y_{ijtk} \le 1 \quad \forall i,j,t \qquad (4.2)$$

$$\sum_i \sum_k y_{ijtk} \leq 1 \quad \forall j, t \tag{4.3}$$

$$\sum_i y_{ijtk} \leq 1 \quad \forall j, t, k \tag{4.4}$$

$$\sum_j \sum_{t=a_i} \sum_k y_{ijtk} \geq 1 \quad \forall i \tag{4.5}$$

$$\sum_j \sum_k y_{ijtk} \leq U \quad \forall i, t \tag{4.6}$$

$$\sum_i BV_{itk} \leq 1 \quad \forall t, k \tag{4.7}$$

$$\sum_j y_{ijtk} \leq M \cdot BV_{itk} \quad \forall i, t, k \tag{4.8}$$

$$\sum_k y_{ijt+1k} - \sum_k y_{ijtk} \leq CR_{ijt} \quad \forall i, j, t \tag{4.9}$$

$$N_{it+1} \leq M \cdot PH_{it} \quad \forall i, t, t \neq T \tag{4.10}$$

$$N_{it} - \sum_j \sum_k R_j \cdot y_{ijtk} = N_{i,t+1} \quad \forall i, t, t \neq T \tag{4.11}$$

$$N_{i,T} \leq 0 \quad \forall i \tag{4.12}$$

$$YH_{it} \leq PH_{it} \quad \forall i, t \tag{4.13}$$

$$YH_{it} \leq TempH_{it} \quad \forall i, t \tag{4.14}$$

$$\sum_j \sum_k y_{ijtk} \geq m \cdot TempH_{it} \quad \forall i,t \qquad (4.15)$$

$$\sum_j \sum_k y_{ijtk} \leq M \cdot TempH_{it} \quad \forall i,t \qquad (4.16)$$

$$YH_{it} \geq m \cdot TempH_{it} \quad \forall i,t \qquad (4.17)$$

$$\sum_j \sum_{t=a_i} y_{ijtk} \geq m \cdot \Delta_{ik} \quad \forall i,k \qquad (4.18)$$

$$\sum_j \sum_{t=a_i} \sum_{k'\neq k} y_{ijtk'} \leq M \cdot (1-\Delta_{ik}) \quad \forall i,k \qquad (4.19)$$

$$\sum_j \sum_{t=a_i} y_{ijtk} \leq M \cdot \Delta_{ik} \quad \forall i,k \qquad (4.20)$$

$$\sum_j \sum_{t=a_i} \sum_{k'\neq k} y_{ijtk'} \geq m \cdot (1-\Delta_{ik}) \quad \forall i,k \qquad (4.21)$$

$$\sum_i \sum_{j'\geq j+1} \sum_{k'\leq k-1} y_{ij'tk'} \leq M \cdot \left(1-\sum_i y_{ijtk}\right) \quad \forall j \leq p,t,k \qquad (4.22)$$

$$y_{ijtk}, \Delta_{ik}, PH_{it}, YH_{it}, TempH_{it}, CR_{ijt}, BV_{i,t,k} \in \{0,1\} \quad \forall i,j,t \qquad (4.23)$$

$$N_{it} \ni \forall i,t \qquad (4.24)$$

The first objective f_1 minimizes the total time the vessels spend at the port. When all the containers are handled, the handling time is calculated by summing the total number of assignments in the time horizon. To calculate the total time, waiting time of the vessels on the bay is also considered. The second objective f_2 minimizes the total number of crane setups.

Constraint set (4.1) ensures that the allocation of a vessel does not exceed the quay length. Constraint set (4.2) implies that a vessel can be assigned to at most one berth. Constraint set (4.3) does not allow any crane to be allocated to more than one vessel at multiple berths at time t. Constraint set (4.4) implies that a single vessel can be served by a certain crane at any given time. Constraint set (4.5) ensures that all arriving vessels are served. Constraint set (4.6) guarantees that the total number of cranes allocated in a time period exceeds the maximum number of cranes that can be allocated to a vessel. By constraint set (4.7), the number of vessels allocated to a berth at a given time is limited to 1. Constraint set (4.8) ensures that the value of BV_{itk} at the considered berth–vessel pair is set to 1 if a vessel is given service at the dock at a given time. In constraint set (4.9), crane setup indicators are updated. By constraint set (4.10), a vessel's PH_{it} value is set to 1, if the vessel has arrived and there are remaining containers. In constraint set (4.11), the number of containers to be handled in each vessel is decreased by the crane handling rate at each period. Constraint set (4.12) ensures that all the containers on the vessel are handled. The logical connection between PH_{it} and YH_{it} is secured by constraint set (4.13). Constraint sets (4.14) through (4.17) formulate the equations for solving the total handling time of each vessel. If an y_{ijtk} assignment exists for a vessel at a given time, the vessel handling time variable, YH_{it}, is set to 1. Constraint sets (4.18) through (4.21) ensure that a vessel is docked at a single berth. Constraint set (4.22) handles the crane passing constraints for static cranes. If a crane j is serving a vessel at berth k, then no other crane with a larger crane id can serve a vessel at any berth that is positioned to its right. In the next section, our solution approach will be discussed.

4.4 Solution Methodology

The solution approach that we propose for solving the integrated BAP and CAP problems with multiobjectives relies on an iterative algorithm consisting of a branch-and-cut solver embedded in the ε-constraint method. ε-Constraint method is a well-recognized technique to solve multicriteria optimization problems (Ehrgott, 2005). Figure 4.1 illustrates our solution algorithm. The ε-constraint method does not aggregate the multiple objectives into one criterion as done in a weighted sum method, but minimizes one of the original objectives and transforms the others into constraints. For bi-objective model, values a and b shown in Figure 4.1 give the range for the objective criteria f_2.

A general multiobjective problem with O objectives may be substituted by the ε-constraint method as follows:

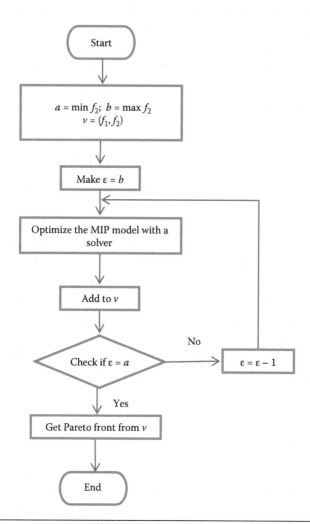

Figure 4.1   Flow chart diagram of the solution algorithm.

$$\min_{x \in X} f_j(x)$$

$$\text{s.t. } f_j(x) \leq \varepsilon_k \quad k = 1,\ldots,O, k \neq j$$

$$\text{where } \varepsilon \in \mathbb{R}^O$$

Here, the choice of the criteria that will be selected to be treated as constraint depends on the problem structure. For our problem, this transformation has been implemented by selecting $f_1(time)$ as objective function and $f_2(setup)$ as constraint. This is mainly due to the highly esteemed customer service levels that are related to time considerations. This transformation is also suitable for the optimization structure as the integer values of the setup parameters allow the parameter ε to be changed by one unit in each subsequent iteration. Additionally, the range for the objective criteria is again appropriate considering the number of iterations that would be required in case of a wider range. Although this range is actually dependent on the instance data examined, nevertheless, the range for time criteria is expected to be broader than of the setup criteria. Therefore, we would anticipate having the number of iterations that needs to be realized as higher in the case where objective $f_1(time)$ was selected to be treated as constraint.

Due to the conflicting nature of both objective criteria, we expect to have the lowest values of one criterion while the other one takes its highest values. This fact allows for the solution algorithm where we may reduce the values of parameter ε for the objective function treated as constraint, to be reduced iteratively, after the selection of one of the criterion as the main objective. To retrieve the interval where parameter ε varies, we solve each problem with a single objective. We expect to have the highest setup cost values when the objective function f_1 is optimized.

4.5 Case Study

Based on the real data obtained from the port of Shanghai container terminal, the model proposed is used to optimize the simultaneous assignment of berths and cranes to the incoming container vessels. The problem has previously been demonstrated by Liang et al. (2009).

Table 4.1 Input for the Computational Study

	SHIP NAME	ARRIVAL TIME	ARRIVAL TIME (IMPLEMENTED)	DUE TIME	TOTAL NUMBER OF LOADING/UNLOADING CONTAINER (TEU)
1	MSG	9:00	10	20:00	428
2	NTD	9:00	10	21:00	455
3	CG	0:30	2	13:00	259
4	NT	21:00	22	23:50	172
5	LZ	0:30	2	23:50	684
6	XY	8:30	10	21:00	356
7	LZI	7:00	8	20:30	435
8	GC	11:30	13	23:50	350
9	LP	21:30	23	23:50	150
10	LYQ	22:00	23	23:50	150
11	CCG	9:00	10	23:50	333

Note that, as the same dataset has later been studied by Han et al. (2010) and Liang et al. (2011), the real case problem might be used as a benchmark. The arrival time, the total number of containers in TEU, and due dates for each vessel are given in Table 4.1.

We represent a 24 h day by 24 equal time intervals and convert all the times in Table 4.1 accordingly. Figure 4.2 illustrates the time scale used for modeling the problem. The same scaling is used for each day.

The berth structure is discrete, and the whole quay area is partitioned into four berths. Since berth lengths are not indicated in the benchmark problem, physical length restrictions are not reflected. There are seven quay cranes, with a handling rate equal to 40 TEUs/h. Due to the lack of available accurate data, the cranes are taken as identical in terms of their handling rates. That, in fact, is a generalization of our model structure, as we allow for variable quay crane handling rate specification. With more realistic crane specifications, our model can be used much more efficiently. The maximum allowable number of cranes assigned to a vessel is 4. In order to show the impact of portable and static cranes, crane ids 6 and 7 are assumed to be portable, that is, move among the berths, while five of the seven cranes are assumed to be static.

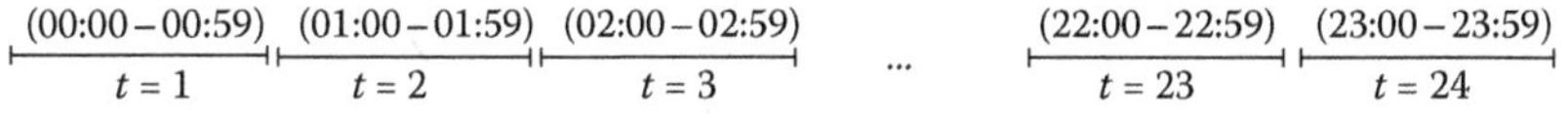

(00:00 – 00:59)	(01:00 – 01:59)	(02:00 – 02:59)	...	(22:00 – 22:59)	(23:00 – 23:59)
$t = 1$	$t = 2$	$t = 3$		$t = 23$	$t = 24$

Figure 4.2 Time implementation frame.

The model is coded in GAMS 22.5 and solved with GUROBI solver for solving integer problems. The preliminary computational experimentation is conducted on NEOS server in January 2012 (Gropp and More, 1997; Czyzyk et al. 1998; Dolan, 2001). The implemented model has 8058 constraints and 3950 variables of which 3749 of them are discrete. The execution of the solver for each instance is reported to have less than 1 CPU s. However, the observed real time is between 5 min (for corner points) and 2 h (for points lying in the center of the Pareto frontier).

The summary results of Pareto solutions are provided in Table 4.2, whereas Figure 4.3 illustrates the optimum Pareto efficient frontier.

Table 4.2 Summary of the Solutions

SOLUTION ID	$F1$: TOTAL SERVICE TIME (H)	$F2$: TOTAL NUMBER OF CRANE SETUP	NONDOMINATED SOLUTION (✓ IF NONDOMINATED)
1	39	42	✗
2	39	41	✓
3	40	38	✗
4	40	37	✗
5	40	36	✓
6	41	35	✓
7	42	34	✗
8	42	33	✓
9	43	32	✓
10	44	31	✓
11	45	30	✓
12	46	29	✓
13	47	28	✓
14	48	27	✓
15	50	26	✓
16	52	25	✓
17	53	24	✓
18	56	23	✓
19	59	22	✓
20	63	21	✓
21	68	20	✓
22	72	19	✓
23	80	18	✓
24	89	18	✗

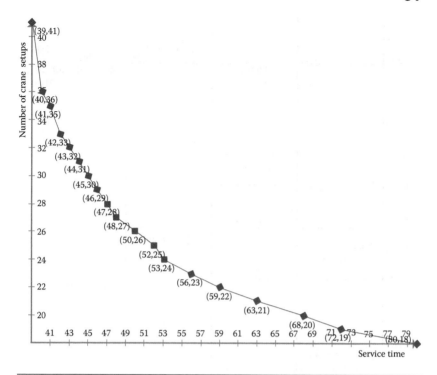

Figure 4.3 Pareto efficient frontier.

Note that not all of our computed solutions contribute to the Pareto efficient frontier, since some of our solutions are dominated by the others. As in the case of solutions #3 and #4, both with total service time equal to 40 min, while total number of crane setups are 38 and 39 respectively, are dominated by solution #5 with exactly same total service time but with less total number of crane setups. Finally, we obtain 19 nondominated solutions out of 24 solutions to form the Pareto efficient frontier. In the study by Liang et al. (2009), the Pareto efficient frontier that is provided has seven solutions. With 19 nondominated solutions, we have further developed the decision support tool by offering an extended number of alternatives to the decision maker.

The second solution in Table 4.2 gives the Pareto optimal solution, which minimizes the service time of the vessels. The relative computational results for the solution are given in Figure 4.4 and Table 4.3. The service time is the difference between the departure time and the arrival time of a vessel. The waiting time is defined as the time

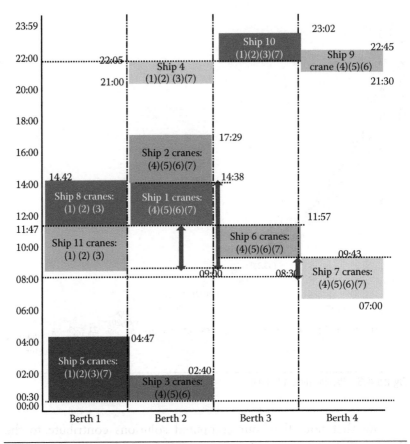

Figure 4.4 The Gantt chart of solution 2.

a vessel spends in the bay before being berthed; that is, the berthing time minus the arrival time. Handling time is the time vessel spends at the port. Delay is the due date minus departure time of the vessel. Note that time scale is converted into minutes for benchmark purposes with earlier studies in literature. In this solution, total service time is 2165, handling time is 1555, waiting time is 610, and delay time is 0. The number of crane setups is 40.

The last nondominated solution in Table 4.2 (solution 23) gives the Pareto optimal solution, which minimizes the quay crane setups. The relative computational results, presented in a similar fashion for this solution, are given in Figure 4.5 and Table 4.4. In solution 23, total service time is 4396, handling time is 3880, waiting time is 516, and delay time is 0. The number of crane setups is 18. Note that this

Table 4.3 Decomposition of the Objectives for Solution 2

	SHIP NAME	ASSIGNED BERTH	WAITING TIME [1] (MIN)	HANDLING TIME [2] (MIN)	SERVICE TIME ([1] + [2])	TOTAL DELAY (MIN)	NUMBER OF CRANE SETUPS
1	MSG	2	177	161	338	0	4
2	NTD	2	338	171	509	0	4
3	CG	2	0	130	130	0	3
4	NT	2	0	65	65	0	4
5	LZ	1	0	257	257	0	4
6	XY	3	73	134	207	0	4
7	LZI	4	0	163	163	0	4
8	GC	1	17	175	250	0	3
9	LP	4	0	75	75	0	3
10	LYQ	3	5	57	62	0	4
11	CCG	1	0	167	167	0	3
Total			610	1555	2165	0	40

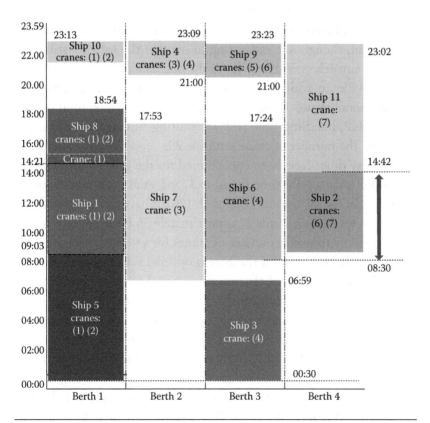

Figure 4.5 The Gantt chart of solution 23.

Table 4.4 Decomposition of the Objectives for Solution 23

	SHIP NAME	ASSIGNED BERTH	WAITING TIME [1] (MIN)	HANDLING TIME [2] (MIN)	SERVICE TIME ([1] + [2])	TOTAL DELAY (MIN)	NUMBER OF CRANE SETUPS
1	MSG	1	3	321	324	0	2
2	NTD	4	0	342	342	0	2
3	CG	3	0	389	389	0	1
4	NT	2	0	129	129	0	2
5	LZ	1	0	513	513	0	2
6	XY	3	0	534	534	0	1
7	LZI	2	0	653	653	0	1
8	GC	1	171	273	444	0	2
9	LP	3	0	113	113	0	2
10	LYQ	1	0	113	113	0	2
11	CCG	4	342	500	842	0	1
Total			516	3880	4396	0	18

solution belongs to an extreme point in the optimal Pareto efficient frontier. It is, therefore, foreseeable to have service time increased by a large extent, though it is interesting to see that the solution still does not cause any delays with respect to the due date given.

To provide an additional example, in Figure 4.6 and Table 4.5, computational results for solution 16 are reported. Here, total service time is 2842, handling time is 2578, waiting time is 264, and delay time is 0. The number of crane setups is 25.

No delay times have been encountered for the given solutions. The minimum total service time found in Liang et al. (2009) is reported as 2165 min for all vessels. Han et al. (2010) have demonstrated the minimum total service time as approximately 36 h for the case where the maximum allowable number of cranes for a vessel is set to 4. Our solution with 2165 min for the Pareto optimal solution, which minimizes the service time of the vessels, is equal to the value found by Liang et al. (2009). Consequently, our study proves the optimality of this solution by implementing an exact algorithm approach.

In both studies by Liang et al. (2009) and Han et al. (2010), crane movements are defined for movements among berths. In our formulation, however, we also take into account the movement among vessels, as the setup cost of a crane to serve a vessel is not neglected (LALB Harbor Safety Committee, 2012). Moreover, in their formulation,

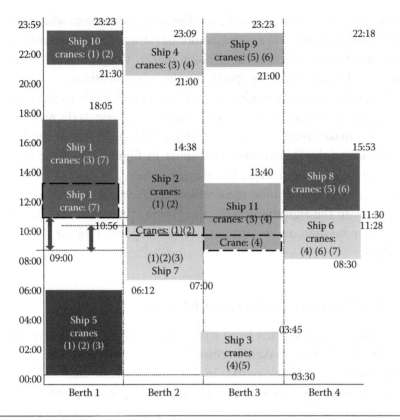

Figure 4.6   The Gantt chart of solution 16.

Table 4.5   Decomposition of the Objectives for Solution 16

	SHIP NAME	ASSIGNED BERTH	WAITING TIME [1] (MIN)	HANDLING TIME [2] (MIN)	SERVICE TIME ([1] + [2])	TOTAL DELAY (MIN)	NUMBER OF CRANE SETUPS
1	MSG	1	148	387	535	0	2
2	NTD	2	116	342	458	0	2
3	CG	3	0	195	195	0	2
4	NT	2	0	129	129	0	2
5	LZ	1	0	342	342	0	3
6	XY	4	0	178	178	0	3
7	LZI	2	0	236	236	0	3
8	GC	4	0	263	263	0	2
9	LP	3	0	113	113	0	2
10	LYQ	1	0	113	113	0	2
11	CCG	3	0	280	280	0	2
Total			264	2578	2842	0	25

cranes are assumed as identical, and the information as to which specific crane is assigned to a vessel cannot be retrieved from the solution, and this decision is left to the decision maker. We, in turn, also support the decision maker by specifying the crane identities.

From the numerical results, one can conclude that when quay crane setup costs are ignored, the total service time of vessels decreases. A decision maker might choose to prefer a solution closer to the left-hand side of the Pareto efficient frontier in Figure 4.3, if the setup costs are not so significant. On the other hand, in case of extreme setup costs of quay cranes, the decision maker is directed toward the solutions in the right-hand side. The Pareto efficient frontier in this case may be used as an efficient decision support tool for decision makers.

4.6 Conclusions and Further Research Directions

Port management is often faced with many challenging problems that require the decision makers to consider numerous issues all at a time. Involvement of multiple parties in the activities associated with container terminal operations makes the port management problem even more complex. The presence of such complications necessitates the use of a decision support tool.

In this study, we propose a decision support tool for the simultaneous berth allocation and crane scheduling problem in consideration of the multiple objectives that need to be satisfied. We first extend the literature by better reflecting practical considerations. We then formulate this problem by bi-objective integer programming. To solve the problem, we follow an ε-constraint method–based solution algorithm to acquire the nondominated berth–crane assignments and schedules as the Pareto optimal frontier. The decision makers may use the obtained optimal Pareto frontier as a decision aid tool. As an insight, we may say that the decisions will be made toward the left-hand side of the frontier if crane setup costs are not so substantial. Conversely, with extreme crane costs, the decision makers are directed toward the solutions in the right-hand side. With this multi-solution approach, decision maker is offered the flexibility of adjusting the balance within conflicting objectives.

We would like to emphasize the fact that this study is part of an ongoing work. We aim to implement our model to other ports of the world to further examine practical considerations that may be required. We will work toward the potential to incorporate our solution procedure with in-house-developed optimization techniques. As a further future work, we believe that the framework we have presented here may further be extended to capture more realistic implementations incorporating issues such as the uncertainty residing in the arrival time of vessels and handling time of cranes. Furthermore, objectives of the model may be analyzed in detail and restructured in parallel to the needs of the decision makers.

As last words, it should be kept in mind that this model is a decision tool that can help decision makers to understand the situation better, rather than finding *the optimum* design. By adjusting parameters or assigning priorities to different objectives, it is possible to obtain a number of satisfactory solutions; however, the ultimate decision always lies with the decision maker.

Acknowledgment

This study is part of a research project funded by TUBITAK (The Scientific and Technological Research Council of Turkey): 1001—The Support Program for Scientific and Technological Research Projects program grant no. 112M865.

References

Bierwirth, C., F. Meisel. 2010. A survey of berth allocation and quay crane scheduling problems in container terminals. *European Journal of Operational Research* 202(3): 615–627.

Carlo, H.J., I.F.A. Vis, K.J. Roodbergen. 2014. Transport operations in container terminals: Literature overview, trends, research directions and classification scheme. *European Journal of Operational Research*, 236(1): 1–13.

Cheong, C.Y., K.C. Tan, D.K. Liu, C.J. Lin. 2010a. Multi-objective and prioritized berth allocation in container ports. *Annals of Operations Research* 180: 63–103.

Cheong, C.Y., M.S. Habibullah, R.S.M. Goh, X. Fu. 2010b. Multi-objective optimization of large scale berth allocation and quay crane assignment problems. In: *Proceedings of the SMC*, Barcelona, Spain, pp. 669–676.

Czyzyk, J., M. Mesnier, J. Moré. 1998. The NEOS server. *IEEE Journal on Computational Science and Engineering* 5: 68–75.

Dolan, E. 2001. The NEOS server 4.0 administrative guide, Technical Memorandum ANL/MCS-TM-250. Mathematics and Computer Science Division, Argonne National Laboratory, Argonne, IL.

Ehrgott, M. 2005. *Multicriteria Optimization.* Springer, Berlin, Germany.

Ehrgott, M., X. Gandibleux. 2002. *Multiple Criteria Optimization: State of the Art Annotated Bibliographic Surveys.* Kluwer Academic, Boston, MA.

Golias, M.M., M. Boilé, S. Theofanis. 2009. Service time based customer differentiation berth scheduling. *Transportation Research Part E: Logistics and Transportation Review* 45(6): 878–892.

Golias, M.M., M. Boilé, S. Theofanis, A.H. Taboada. 2010. A multi-objective decision and analysis approach for the berth scheduling problem. *International Journal of Information Technology Project Management* 1(1): 54–73.

Gropp, W., J. Moré. 1997. Optimization environments and the NEOS server. In: M.D. Buhmann and A. Iserles (eds.), *Approximation Theory and Optimization.* Cambridge University Press, Cambridge, U.K., pp. 167–182.

Han, X., Z. Lu, L. Xi. 2010. A proactive approach for simultaneous berth and quay crane scheduling problem with stochastic arrival and handling time. *European Journal of Operational Research* 207: 1327–1340.

Imai, A., J.-T. Zhang, E. Nishimura, S. Papadimitriou. 2007. The berth allocation problem with service time and delay time objectives. *Maritime Economics & Logistics* 9: 269–290.

LALB Harbor Safety Committee. 2012. Vessel terminal gantry crane safety. http://www.yjcrane.com/solution/gantry-crane/vessel-terminal-gantry-crane-safety.html (last accessed February 16, 2012).

Liang, C., J. Guo, Y. Yang. 2011. Multi-objective hybrid genetic algorithm for quay crane dynamic assignment in berth allocation planning. *Journal of Intelligent Manufacturing* 22: 471–479.

Liang, C., Y. Huang, Y. Yang. 2009. A quay crane dynamic scheduling problem by hybrid evolutionary algorithm for berth allocation planning. *Computers & Industrial Engineering* 56(3): 1021–1028.

Rodrigue, J.P., T. Notteboom, B. Slack. 2009. Transportation modes. In: Rodrigue, J.P., C. Comtois, B. Slack (eds.), *The Geography of Transport Systems.* Routledge, New York, Chapter 3.

UNCTAD/RMT/2013. 2013. Review of maritime transport. United Nations Publications, Geneva, Switzerland.

Zhang, C., L. Zheng, Z. Zhang, L. Shi, A.J. Armstrong. 2010. The allocation of berths and quay cranes by using a sub-gradient optimization technique. *Computers & Industrial Engineering* 58: 40–50.

Zhou, P., H. Kang. 2008. Study on berth and quay-crane allocation under stochastic environments in container terminal. *Systems Engineering—Theory & Practice* 28: 161–169.

5

ROUTE SELECTION PROBLEM IN THE ARCTIC REGION FOR THE GLOBAL LOGISTICS INDUSTRY

BEKIR SAHIN

Contents

5.1 Introduction

Logistics is a process of distribution network management and optimization of the flow of resources. Therefore, logistics management benefits from using optimal product transportation. The transportation locations of the economic world are undergoing a dramatic change with the emergence of new Arctic seaways (Wilson et al. 2004). The melting of sea ice in the northern hemisphere is being observed with great attention. As a result of both greenhouse effects and seasonal fluctuations of long-term average temperatures, a historical opportunity presents itself to extend maritime transport over the Arctic region. For instance, the Northern Sea route shortens the Yokohama–London distance via the Suez Canal from 11.447 to 7.474

Figure 5.1 Overview on the Northern Sea Route and the Suez Route. (Adapted from Schøyen, H. and Bråthen, S., *J. Transp. Geogr.*, 19, 977, 2011.)

nautical miles. Figure 5.1 illustrates the Arctic Sea routes that shorten the traditional routes.

The possibility of a shipping route over the Arctic region will significantly lessen time and energy spent on long transportations on a regular basis. However, because the possible shipping routes are still covered by floating ice (i.e., open ice, closed ice), the Arctic routing is facing an ongoing debate with its highly technical circumstances. Navigational track (route) optimization, entry into the ice field, and route selection are some challenges in this field. Among these debates, route selection is the main concern of this chapter and it depends on a number of factors such as the dimensions and the physical conditions of the route.

Arctic navigation is a new concept with a short literature including track optimization among other aspects (Thomson and Sykes 1988; Ari et al. 2013). Once a navigational route is selected, various studies can improve the navigational quality in terms of time, structural stress, and fuel consumption. However, route selection is the primary problem, which is not discussed in earlier studies.

The route selection problem can be categorized into static route selection and dynamic route selection. The static route selection approach is based on instant inputs of indicators while assuming that ice field and weather conditions are stationary over the intended navigational sea field. In the case of the dynamic approach, the ice field and weather may have variations over the region and the size and direction of the vectors may change over time.

As an introduction to the problem, this chapter deals with the static route selection approach from the perspective of subjective judgments of shipmasters. It is certain that these field experts have enough experience and knowledge on ice navigation operations and winterization. The problem is investigated by using the fuzzy analytical hierarchy process (F-AHP). The reasons behind the selection of F-AHP are twofold: first, AHP is very useful for handling both quantitative and subjective matters and, second, fuzzy extension facilitates the process for subjects in the survey by using linguistic representations. Decision makers' (DMs') uncertainty is a common case and, based on the drawbacks of uncertainty, fuzzy transformations help the moderator (i.e., researcher) to collect a span of data rather than a single crisp number with an unknown degree of certainty.

5.2 Route Selection Problem in the Arctic Region

Logistics activities in the Arctic region are regularly conducted by ferries, big roll-on/roll-off (RO-Ros), and icebreaker convoys. These powerful vessels leave their tracks. Therefore, recent tracks are preferable for navigation in ice-covered sea regions. Figure 5.2 is an empirical image which shows an objective vessel and previous tracks opened by icebreakers or other vessels. Ice navigation becomes hard in such an environment, and route selection management requires field experience.

For vessels traveling from one point to another in ice, it is important to detect the optimal routes that reduce travel time, fuel consumption, and getting stuck in ice. Seafarers gather route information from various sources such as radar (e.g., automatic radar plotting aids, ARPA), satellite images, infrared cameras, visual recognition, and charts. After many continuous observations, available paths are drawn as shown in Figure 5.3. There are three different possible routes connecting the starting point to the final destination.

The average route width (ARW), slot availability (S), maximum width along with the track (Max), minimum width along with the track (Min), ice concentration (IC), route length (RL), sea depth (SD),

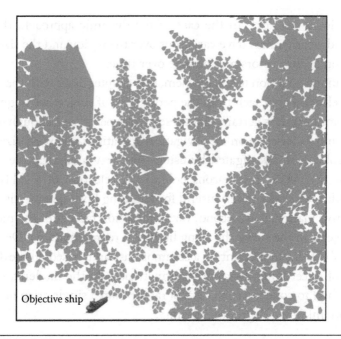

Figure 5.2 A ship prepares to navigate in ice-covered sea regions.

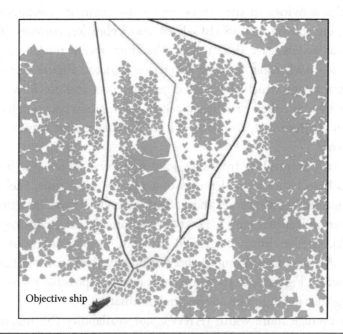

Figure 5.3 Routes for the ship navigation in ice-covered sea regions (tracks 1, 2, and 3 from left to right).

Ice concentration		
Open water	<1/10	
Very open drift	1–3/10	
Open drift	4–6/10	
Close pack	7–8/10	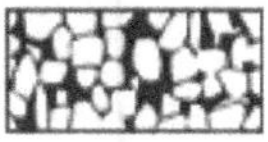
Very close pack	9/10	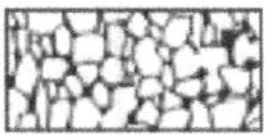
Very close pack	9+/10	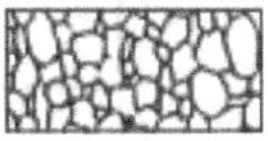
Compact/consolidated ice	10/10	

Figure 5.4 Ice concentration diagram.

and sharp bend (SB) are the eight selective parameters that affect ice navigation based on time, oil consumption, expenditure, and safety. We assume that this platform is static and that floes are constant. In this approach, IC (Figure 5.4), which is calculated at a ratio of one-tenth, roughly corresponds to an average ice concentration on the selected tracks.

In a narrow and straight track, if a ship meets a big piece of fast ice that is broken off and acts like a fender, there is a great risk of collision (Buysee 2007). In this research, all floes are static and we assume that there are no obstacles in the tracks. However, when two vessels are in a crossing situation, the safest way is to stop just outside the track, which we call a "slot" (see Figure 5.5). If there is an available slot, the safer method is to use the slot for clearing the track and the vessel drifts into slot till the track is cleared. Once the vessel has passed, one can easily get unstuck by an astern maneuver.

Figure 5.5   Track meet of two vessels and slot availability.

Sea depth is crucial for the vessels' keels, hulls, and propellers. Hard ice may damage the vessel physically and stop the maneuverability of the vessel. Engaging an SB is another navigational challenge in ice-covered waters. This maneuvering technique requires special skill and experience (see Figure 5.6). The motor vessel's speed should be reduced to half 5 cables ahead (185 × 5 m), and the vessel should be steered to the left and then given the command of full astern.

5.3 Methodology

5.3.1 Linguistic Variable

In a natural or artificial language, a linguistic variable is, for example, weather, and the values are expressed as fuzzy words or sentences such as hot, very hot, cold, and very cold instead of numbers (Bellman and Zadeh 1977).

A linguistic variable has an approximation character, which is either too complex or unclear to be described in quantitative terms. Linguistic variables are commonly applied in humanistic systems such

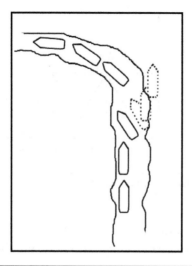

Figure 5.6 Engaging a sharp bend.

as human decision processes, artificial intelligence, pattern recognition, law, medical realms, economy, and related areas (Zadeh 1975).

5.3.2 Fuzzy Sets and Triangular Fuzzy Numbers

A fuzzy set was first developed by Zadeh (1965) and introduced by Bellman and Zadeh (1977). A triangular fuzzy number is a convex and normalized fuzzy set $\tilde{A}$ and $\mu_{\tilde{A}}(x)$ is the continuous linear function, which is a membership function of $\tilde{A}$.

The definition of a triangular fuzzy number $\tilde{A} = (l, m, u)$ is

$$\mu_{\tilde{A}}(x) = \begin{cases} 0, & x < l, \\ \dfrac{(x-l)}{(m-l)}, & l \leq x < m, \\ 1, & x = m, \\ \dfrac{(u-x)}{(u-m)}, & m < x \leq u, \\ 0, & u < x. \end{cases} \tag{5.1}$$

where

l and u are, respectively, the lower (smallest possible value) and upper (most promising value) bounds of the fuzzy number $\tilde{A}$ and m is the midpoint

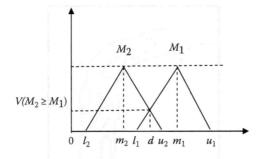

Figure 5.7   The intersection between M_1 and M_2.

A triangular fuzzy number is shown in Figure 5.7.

Consider two positive triangular fuzzy numbers (l_1, m_1, u_1) and (l_2, m_2, u_2); then

$$\left(l_1, m_1, u_1\right) + \left(l_2, m_2, u_2\right) = \left(l_1 + l_2, m_1 + m_2, u_1 + u_2\right)$$

$$\left(l_1, m_1, u_1\right) \cdot \left(l_2, m_2, u_2\right) = \left(l_1 \cdot l_2, m_1 \cdot m_2, u_1 \cdot u_2\right)$$

$$\left(l_1, m_1, u_1\right)^{-1} \approx \left(\frac{1}{u_1}, \frac{1}{m_1}, \frac{1}{l_1}\right)$$

$$\left(l_1, m_1, u_1\right) \cdot k = \left(l_1 \cdot k, m_1 \cdot k, u_1 \cdot k\right)$$

where k is a positive number.

The vertex method is used to calculate the distance between two triangular fuzzy numbers (Chen 2000):

$$d_v\left(\tilde{m}, \tilde{n}\right) = \sqrt{\frac{1}{3}\left[\left(l_1 - l_2\right)^2 + \left(m_1 - m_2\right)^2 + \left(u_1 - u_2\right)^2\right]}$$

5.3.3 Fuzzy Analytic Hierarchy Process

Saaty (1980) proposed the first AHP as a decision-making tool. This method is widely used by researchers (Weck et al. 1997; Lee et al. 1999; Leung and Cao 2000). The main purpose of AHP is to use the experts' knowledge; however, the classical AHP does not reflect the human thinking style (Chang 1996) because it uses the exact values when comparing the criteria with alternatives (Cakir and Canpolat 2008). There has been a lot of criticism regarding the classical AHP because of its unbalanced scale, uncertainty, and imprecision of pairwise comparisons (Kahraman et al. 2004). F-AHP is more accurate and has been developed to handle these shortcomings. Laarhoven and Pedrycz (1983) proposed the first F-AHP by the comparisons of fuzzy ratios. Buckley (1985) worked on trapezoidal fuzzy numbers to evaluate the alternatives with respect to the criteria. For the pairwise comparisons, Chang (1996) used the extent analysis method to calculate the synthetic extent values.

The steps of the extent synthesis method are as follows.

Let $X = \{x_1, x_2, \ldots, x_n\}$ be an object set and $G = \{g_1, g_2, \ldots, g_n\}$ be a goal set. Each object is taken, and an extent analysis is performed for each goal. Therefore, m extent analysis values for each object can be obtained:

$$M_{gi}^1, M_{gi}^2, \ldots, M_{gi}^m, \quad i = 1, 2, \ldots, n \tag{5.2}$$

where all M_g^j ($j = 1, 2, \ldots, m$) are triangular fuzzy numbers.

Step 1: For the ith object, the value of the fuzzy synthetic extent is defined as

$$S_i \sum_{j=1}^m M_{gi}^j \otimes \left[\sum_{i=1}^n \sum_{j=1}^m M_{gi}^j \right]^{-1} \tag{5.3}$$

Obtaining $\sum_{j=1}^m M_{gi}^j$ the fuzzy addition operation of m extent analysis values for a particular matrix is performed:

$$\sum_{j=1}^m M_{gi}^j = \left(\sum_{j=1}^m l_j, \sum_{j=1}^m m_j, \sum_{j=1}^m u_j \right) \tag{5.4}$$

The fuzzy addition operation of M_{gi}^j ($j = 1, 2, \ldots, m$) values is performed:

$$\sum_{i=1}^n \sum_{j=1}^m M_{gi}^j = \left(\sum_{j=1}^m l_j, \sum_{j=1}^m m_j, \sum_{j=1}^m u_j \right) \tag{5.5}$$

The inverse of the vector in Equation 5.3 is computed:

$$\left[\sum_{i=1}^{n}\sum_{j=1}^{m}M_{gi}^{j}\right]^{-1}=\left(\frac{1}{\sum_{i=1}^{n}u_i},\frac{1}{\sum_{i=1}^{n}m_i},\frac{1}{\sum_{i=1}^{n}l_i}\right) \quad (5.6)$$

Step 2: The height of a fuzzy set $hgt(A)$ is the maximum of the membership grades of A, $hgt(A)=\sup_{x\in X}\mu_A(x)$.
The degree of possibility of $M_2=(l_2,\ m_2,\ u_2)\ge M_1=(l_1,\ m_1,\ u_1)$ is defined as follows:

$$V\left(M_2\ge M_1\right)=\sup_{y\ge x}\left[\min\left(\mu_{M_1}(x),\mu_{M_2}(y)\right)\right] \quad (5.7)$$

and can also be expressed as

$$V\left(M_2\ge M_1\right)=hgt\left(M_1\cap M_2\right)$$

$$=\mu_{M_2}(d)=\begin{cases}1, & \text{if } m_2\ge m_1\\ 0, & \text{if } l_1\ge u_2\\ \dfrac{l_1-u_2}{(m_2-u_2)-(m_1-l_1)}, & \text{otherwise}\end{cases}$$

$$(5.8)$$

Figure 5.8 illustrates that d is the y-axis value of the highest intersection point D between μ_{M_1} and μ_{M_2}.
Both $V(M_1\ge M_2)$ and $V(M_2\ge M_1)$ should be known for the comparison of M_1 and M_2.

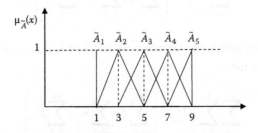

Figure 5.8 Fuzzy number of linguistic variable set.

Step 3: The degree of possibility for a convex fuzzy number to be greater than k convex fuzzy numbers M_i $(i=1, 2, ...)$ can be defined by

$$V(M \geq M_1, M_2, ..., M_k) = V[(M \geq M_1), (M \geq M_2), ...,$$
$$(M \geq M_k)] = \min V(M \geq M_i), i = 1, 2, 3, ..., k. \qquad (5.9)$$

Assume that $d(A_i) = \min V(S_i \geq S_k)$ for $k = 1, 2, ..., n; k \neq i$. Then the weight vector is given by

$$W' = \left(d'(A_1), d'(A_2), ..., d'(A_n) \right)^T \qquad (5.10)$$

where A_i $(i = 1, 2, ..., n)$ are n elements.

Step 4: Normalization and normalized weight vectors are

$$W = \left(d(A_1), d(A_2), ..., d(A_n) \right)^T \qquad (5.11)$$

where W is a nonfuzzy number.

The nonnumerical values are expressed as fuzzy linguistic variables, which help the DM to describe the pairwise comparison of each criterion with its alternative, as reflected in Saaty's (1977) nine-point fundamental scale (see Figure 5.9).

The assigned linguistic comparison terms (Chiclana 1998; Chan et al. 2000; Cakir and Canpolat 2008; Gumus 2009) and their equivalent fuzzy numbers considered in this chapter are given in Table 5.1.

For solving the current problem, an individual aggregation matrix is conducted by expert prioritization, which is called the lambda coefficient.

Let $A = (a_{ij})n \times n$, where $a_{ij} > 0$ and $a_{ij} \times a_{ji} = 1$, be a judgment matrix. The prioritization method denotes the process of acquiring a priority vector. $w = (w_1, w_2, ..., w_n)^T$ where $w_i \geq 0$ and $\sum_{i=1}^{n} w_i = 1$, from the judgment matrix A.

Let $D = \{d_1, d_2, ..., d_m\}$ be the set of experts, and $\lambda = \{\lambda_1, \lambda_2, ..., \lambda_m\}$ be the weight vector of the DMs, where $\lambda_k > 0$, $k = 1, 2, ..., m$, and $\sum_{k-1}^{m} \lambda_k = 1$.

Let $E = \{e_1, e_2, ..., e_m\}$ be the set of the experience in the professional career (in years for this chapter) for each expert, and λ_k for each expert is defined by

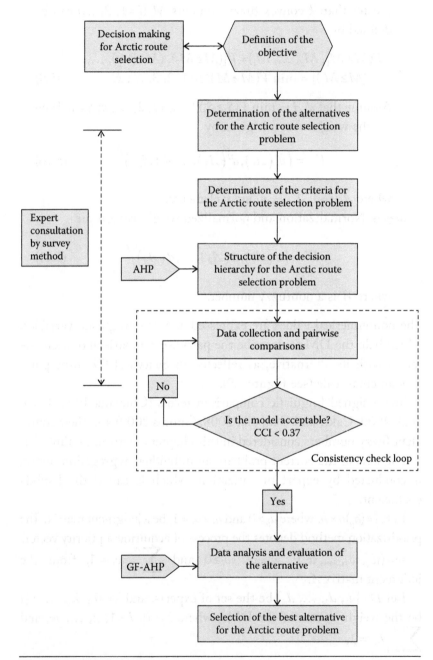

Figure 5.9 GF-AHP procedure.

Table 5.1 Membership Function of Linguistic Scale

FUZZY NUMBER	LINGUISTIC SCALES	MEMBERSHIP FUNCTION	INVERSE
$\tilde{A}_1$	Equally important	(1, 1, 1)	(1, 1, 1)
$\tilde{A}_2$	Moderately important	(1, 3, 5)	(1/5, 1/3, 1)
$\tilde{A}_3$	More important	(3, 5, 7)	(1/7, 1/5, 1/3)
$\tilde{A}_4$	Strongly important	(5, 7, 9)	(1/9, 1/7, 1/5)
$\tilde{A}_5$	Extremely important	(7, 9, 9)	(1/9, 1/9, 1/7)

$$\lambda_k = \frac{e_k}{\sum_{k=1}^{m} e_k} \tag{5.12}$$

Let $A(k) = (a_{ij}^{(k)})_{n \times n}$ be the judgment matrix that is gathered by the DM *dk*.

$w_i^{(k)}$ is the priority vector of criteria for each expert calculated by

$$w_i^{(k)} = \frac{\left(\prod_{j=1}^{n} a_{ij} \right)^{1/n}}{\sum_{i=1}^{n} \left(\prod_{j=1}^{n} a_{ij} \right)^{1/n}} \tag{5.13}$$

The individual priority aggregation is defined by

$$w_i^{(w)} = \frac{\prod_{k=1}^{m} \left(w_i^{(k)} \right)^{\lambda_k}}{\sum_{i=1}^{n} \prod_{k=1}^{m} \left(w_i^{(k)} \right)^{\lambda_k}} \tag{5.14}$$

where $w_i^{(w)}$ is the aggregated weight vector. Then the extent synthesis method (Chang 1996) is applied for the consequent selection. A pairwise comparison between the alternatives *i* and *j* for criterion *C* is defined by

$$a_{ij}^{C} = \frac{A_r^i}{A_r^j} \tag{5.15}$$

where A_r^i is the rank valuation set of alternative *i*. By the final consistency control, the procedure of generic fuzzy AHP (GF-AHP) is achieved. Consistency control and centric consistency index (CCI) for F-AHP applications are described in the following section.

5.3.4 Centric Consistency Index

According to Saaty's approach, all DMs' matrix should be consistent to analyze the selection problem (Saaty and Vargas 1987). For the consistency control of the F-AHP method, Duru et al. (2012) proposed a CCI based on the geometric consistency index (Crawford and Williams 1985; Aguarón and Moreno-Jimenez 2003). The calculation of the CCI algorithm is as follows:

$$CCI(A) = \frac{2}{(n-1)(n-2)} \sum_{i<j} \left(\log \frac{a_{Lij} + a_{Mij} + a_{Uij}}{3} \right.$$

$$\left. - \log \frac{W_{Li} + W_{Mi} + W_{Ui}}{3} + \log \frac{W_{Lj} + W_{Mj} + W_{Uj}}{3} \right)^2 \quad (5.16)$$

When $CCI(A)$ is 0, A is fully consistent. Aguarón also expresses the thresholds $(\overline{GCI})$ as $(\overline{GCI}) = 0.31$ for $n = 3$, $(\overline{GCI}) = 0.35$ for $n = 4$, and $(\overline{GCI}) = 0.37$ for $n > 4$. When $CCI(A) < (\overline{GCI})$, it means that this matrix is sufficiently consistent.

5.4 GF-AHP Design and Application for Track Selection

GF-AHP is a novel extended form of conventional methods of F-AHP (Bulut et al. 2012). GF-AHP is applied to our intended problem because of its many novel contributions over traditional AHP methods. First, GF-AHP is able to execute uncertain consultations. Second, GF-AHP proposes a DM weighting algorithm to combine with the F-AHP. Third, GF-AHP is improved for direct numerical inputs. Fourth, GF-AHP proposes a consistency check index method. The GF-AHP procedure consists of these superior qualities.

Arrangement of the hierarchy is important when using GF-AHP. The determination of the objectives, criteria, and alternatives are placed in a hierarchical structure (Figure 5.10). By a presurvey method, all criteria are determined. The main criteria such as the ARW, S, Max, Min, IC, RL, SD, and SB are analyzed for their impact on the

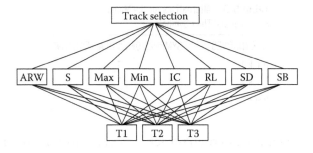

Figure 5.10 Hierarchy of the track selection process.

alternatives. Then the survey is applied to 12 DM. Seven of them are master mariners, three of them are company representatives, and two of them are academicians. Figure 5.10 shows the corresponding hierarchical design of the track selection problem.

The F-AHP approach is employed to determine the weights of criteria and alternative components of the decision hierarchies. In the application of the AHP approach, a pairwise comparison table is formed.

As a first step in the application of the F-AHP approach, for each criterion, weights and priorities are compared pairwise using a fuzzy extension of Saaty's 1–9 scale (Table 5.1).

The F-AHP approach is also applied to calculate weights of alternatives for each criterion (Table 5.2). Table 5.3 shows the alternatives for the track selection and their symbols.

For the decision-making process, a pairwise comparison survey was conducted and reported as follows. The individual fuzzy judgment

Table 5.2 Criteria for the Model of Track Selection and Their Symbols

CRITERIA	SYMBOLS OF EACH CRITERION
Average route width	ARW
Slot availability	S
Maximum width alongside the track	Max
Minimum width alongside the track	Min
Ice concentration	IC
Route length	RL
Sea depth	SD
Sharp bend	SB

Table 5.3 Alternatives for the Model of Track Selection and Their Symbols

ALTERNATIVES	SYMBOLS OF EACH ALTERNATIVE
Track 1	T_1
Track 2	T_2
Track 3	T_3

matrices, which assess criteria-to-criteria comparison, are presented in Table 5.4. Lambda (λ) is the expertise priority of the DM based on the time spent for this industry.

Table 5.5 shows the individual fuzzy priority vector and aggregated weight vector. Each DM usually finds that ARW is the most important factor with its 0.34 value (midpoint) and S is the second crucial indicator (0.27). Other aggregated weight coefficients that contribute to the final outcome are 0.18, 0.17, 0.11, 0.10, and 0.07.

The mean aggregated weight (MAW) is calculated for each criterion as 0.27, 0.22, 0.09, 0.16, 0.10, 0.15, 0.10, and 0.07, respectively (Table 5.6). The aggregated weight vector is computed by the expert priority vector of DMs (λ) and the individual priority vector of each DM. The result is consistent as the CCI is 0.01 less than the threshold of 0.37.

The extent synthesis is performed for the Arctic route selection problem as follows:

$S_{ARW} = (8.81, 19.81, 30.01) \otimes (1/45.84, 1/86.31, 1/132.89) = (0.07, 0.23, 0.65)$

$S_S = (7.68, 17.28, 26.41) \otimes (1/52.64, 1/100.22, 1/153.81) = (0.06, 0.20, 0.58)$

$S_{Max} = (4.20, 5.82, 8.88) \otimes (1/52.64, 1/100.22, 1/153.81) = (0.03, 0.07, 0.19)$

$S_{Min} = (6.77, 13.24, 20.38) \otimes (1/52.64, 1/100.22, 1/153.81) = (0.05, 0.15, 0.44)$

$S_{IC} = (4.71, 6.97, 11.27) \otimes (1/52.64, 1/100.22, 1/153.81) = (0.04, 0.08, 0.25)$

$S_{RL} = (6.49, 11.26, 16.40) \otimes (1/52.64, 1/100.22, 1/153.81) = (0.05, 0.13, 0.36)$

$S_{SD} = (4.04, 7.59, 10.93) \otimes (1/52.64, 1/100.22, 1/153.81) = (0.03, 0.09, 0.24)$

Table 5.4 Individual Fuzzy Judgment Matrix for Criteria of Track Selection

		ARW	S	MAX	MIN	IC	RL	SD	SB
DM₁ λ=0.04	ARW	(1,1,1)	(3,5,7)	(5,7,9)	(3,5,7)	(1,3,5)	(1/5,1/3,1)	(1,3,5)	(1,3,5)
	S	(1/7,1/5,1/3)	(1,1,1)	(3,5,7)	(1,3,5)	(1,3,5)	(1,1,1)	(1,3,5)	(1,3,5)
	Max	(1/9,1/7,1/5)	(1/7,1/5,1/3)	(1,1,1)	(1/5,1/3,1)	(1/7,1/5,1/3)	(1/5,1/3,1)	(1,3,5)	(1,3,5)
	Min	(1/5,1/3,1)	(1/5,1/3,1)	(1,3,5)	(1,1,1)	(1,1,1)	(1,1,1)	(1,3,5)	(1,3,5)
	IC	(1/5,1/3,1)	(1/5,1/3,1)	(3,5,7)	(1,1,1)	(1,3,5)	(1/5,1/3,1)	(1,3,5)	(1,3,5)
	RL	(1,3,5)	(1,1,1)	(1,3,5)	(1,1,1)	(1,3,5)	(1,1,1)	(1,3,5)	(1,3,5)
	SD	(1/5,1/3,1)	(1/5,1/3,1)	(1/5,1/3,1)	(1/5,1/3,1)	(1/5,1/3,1)	(1,1,1)	(1,1,1)	(1,3,5)
	SB	(1/5,1/3,1)	(1/5,1/3,1)	(1/5,1/3,1)	(1/5,1/3,1)	(1/5,1/3,1)	(1,1,1)	(1/5,1/3,1)	(1,1,1)
DM₂ λ=0.16	ARW	(1,1,1)	(1,3,5)	(1,1,1)	(1,3,5)	(1,3,5)	(1,1,1)	(1,1,1)	(1,3,5)
	S	(1/5,1/3,1)	(1,1,1)	(1,3,5)	(1,3,5)	(1,3,5)	(1,1,1)	(1,1,1)	(1,3,5)
	Max	(1,1,1)	(1/5,1/3,1)	(1,1,1)	(1/7,1/5,1/3)	(1,1,1)	(1,1,1)	(1,1,1)	(1,3,5)
	Min	(1/5,1/3,1)	(1/5,1/3,1)	(3,5,7)	(1,1,1)	(1,3,5)	(1,1,1)	(1,1,1)	(1,3,5)
	IC	(1/5,1/3,1)	(1/5,1/3,1)	(1,1,1)	(1/5,1/3,1)	(1,1,1)	(1,1,1)	(1,1,1)	(1,3,5)
	RL	(1,1,1)	(1,1,1)	(1,1,1)	(1,1,1)	(1,1,1)	(1,1,1)	(1,1,1)	(1,3,5)
	SD	(1,1,1)	(1,1,1)	(1,1,1)	(1,1,1)	(1,1,1)	(1,1,1)	(1,1,1)	(1,3,5)
	SB	(1/5,1/3,1)	(1/5,1/3,1)	(1/5,1/3,1)	(1/5,1/3,1)	(1/5,1/3,1)	(1,1,1)	(1,1,1)	(1,1,1)
DM₃ λ=0.08	ARW	(1,1,1)	(1,3,5)	(3,5,7)	(5,7,9)	(1,3,5)	(1,3,5)	(1,3,5)	(1,3,5)
	S	(1/5,1/3,1)	(1,1,1)	(1,1,1)	(1,1,1)	(1,3,5)	(1,1,1)	(1,3,5)	(1,3,5)
	Max	(1/7,1/5,1/3)	(1,1,1)	(1,1,1)	(1,1,1)	(1,3,5)	(1,3,5)	(1,3,5)	(1,3,5)
	Min	(1/9,1/7,1/5)	(1,1,1)	(1,1,1)	(1,1,1)	(1,3,5)	(1,3,5)	(1,3,5)	(1,3,5)
	IC	(1/5,1/3,1)	(1/5,1/3,1)	(1/5,1/3,1)	(1/5,1/3,1)	(1,1,1)	(1,1,1)	(1,3,5)	(1,3,5)
	RL	(1/5,1/3,1)	(1,1,1)	(1/5,1/3,1)	(1/5,1/3,1)	(1,1,1)	(1,1,1)	(1,3,5)	(1,3,5)
	SD	(1/5,1/3,1)	(1/5,1/3,1)	(1/5,1/3,1)	(1/5,1/3,1)	(1/5,1/3,1)	(1/5,1/3,1)	(1,1,1)	(1,3,5)
	SB	(1/5,1/3,1)	(1/5,1/3,1)	(1/5,1/3,1)	(1/5,1/3,1)	(1/5,1/3,1)	(1/5,1/3,1)	(1,1,1)	(1,1,1)

(Continued)

Table 5.4 (Continued) Individual Fuzzy Judgment Matrix for Criteria of Track Selection

		ARW	S	MAX	MIN	IC	RL	SD	SB
DM₄ λ=0.09	ARW	(1, 1, 1)	(1, 3, 5)	(3, 5, 7)	(1, 3, 5)	(1, 3, 5)	(1, 3, 5)	(1, 3, 5)	(1, 3, 5)
	S	(1/5, 1/3, 1)	(1, 1, 1)	(1, 3, 5)	(1, 3, 5)	(1, 3, 5)	(3, 5, 7)	(1, 3, 5)	(1, 3, 5)
	Max	(1/7, 1/5, 1/3)	(1/5, 1/3, 1)	(1, 1, 1)	(1, 3, 5)	(1, 3, 5)	(1, 1, 1)	(1, 3, 5)	(1, 3, 5)
	Min	(1/5, 1/3, 1)	(1/5, 1/3, 1)	(1/5, 1/3, 1)	(1, 1, 1)	(1, 3, 5)	(1, 1, 1)	(1, 3, 5)	(1, 3, 5)
	IC	(1/5, 1/3, 1)	(1/5, 1/3, 1)	(1/5, 1/3, 1)	(1/5, 1/3, 1)	(1, 1, 1)	(1, 1, 1)	(1, 3, 5)	(1, 3, 5)
	RL	(1/5, 1/3, 1)	(1/7, 1/5, 1/3)	(1, 1, 1)	(1, 1, 1)	(1, 1, 1)	(1, 1, 1)	(1, 3, 5)	(1, 3, 5)
	SD	(1/5, 1/3, 1)	(1/5, 1/3, 1)	(1/5, 1/3, 1)	(1/5, 1/3, 1)	(1/5, 1/3, 1)	(1/5, 1/3, 1)	(1, 1, 1)	(1, 3, 5)
	SB	(1/5, 1/3, 1)	(1/5, 1/3, 1)	(1/5, 1/3, 1)	(1/5, 1/3, 1)	(1/5, 1/3, 1)	(1/5, 1/3, 1)	(1/5, 1/3, 1)	(1, 1, 1)
DM₅ λ=0.09	ARW	(1, 1, 1)	(1, 3, 5)	(1, 3, 5)	(1, 3, 5)	(1, 3, 5)	(3, 5, 7)	(1, 3, 5)	(1, 3, 5)
	S	(1/5, 1/3, 1)	(1, 1, 1)	(1, 3, 5)	(1, 3, 5)	(1, 3, 5)	(1, 1, 1)	(1, 3, 5)	(1, 3, 5)
	Max	(1/5, 1/3, 1)	(1/5, 1/3, 1)	(1, 1, 1)	(1, 1, 1)	(1, 1, 1)	(1, 1, 1)	(1, 1, 1)	(1, 1, 1)
	Min	(1/5, 1/3, 1)	(1/5, 1/3, 1)	(1, 1, 1)	(1, 1, 1)	(1, 3, 5)	(1, 1, 1)	(1, 3, 5)	(1, 3, 5)
	IC	(1/5, 1/3, 1)	(1/5, 1/3, 1)	(1, 1, 1)	(1/5, 1/3, 1)	(1, 1, 1)	(1/5, 1/3, 1)	(1, 1, 1)	(1/5, 1/3, 1)
	RL	(1/7, 1/5, 1/3)	(1, 1, 1)	(1, 1, 1)	(1, 1, 1)	(1, 3, 5)	(1, 1, 1)	(1, 3, 5)	(1, 3, 5)
	SD	(1/5, 1/3, 1)	(1/5, 1/3, 1)	(1, 1, 1)	(1/5, 1/3, 1)	(1, 1, 1)	(1/5, 1/3, 1)	(1, 1, 1)	(1, 3, 5)
	SB	(1/5, 1/3, 1)	(1/5, 1/3, 1)	(1, 1, 1)	(1/5, 1/3, 1)	(1, 3, 5)	(1/5, 1/3, 1)	(1/5, 1/3, 1)	(1, 1, 1)
DM₆ λ=0.05	ARW	(1, 1, 1)	(3, 5, 7)	(1, 3, 5)	(1, 3, 5)	(1, 3, 5)	(1, 3, 5)	(1, 3, 5)	(1, 3, 5)
	S	(1/7, 1/5, 1/3)	(1, 1, 1)	(1, 1, 1)	(1, 1, 1)	(1, 3, 5)	(1, 3, 5)	(1, 3, 5)	(1, 3, 5)
	Max	(1/5, 1/3, 1)	(1/5, 1/3, 1)	(1, 1, 1)	(1, 1, 1)	(1, 1, 1)	(1, 1, 1)	(1, 1, 1)	(1, 1, 1)
	Min	(1/5, 1/3, 1)	(1/5, 1/3, 1)	(1, 1, 1)	(1, 1, 1)	(1, 1, 1)	(1, 1, 1)	(1, 3, 5)	(1, 3, 5)
	IC	(1/5, 1/3, 1)	(1/5, 1/3, 1)	(1, 1, 1)	(1/5, 1/3, 1)	(1, 1, 1)	(1, 1, 1)	(1, 1, 1)	(1, 1, 1)
	RL	(1/5, 1/3, 1)	(1/5, 1/3, 1)	(1, 1, 1)	(1, 1, 1)	(1, 1, 1)	(1, 1, 1)	(1, 3, 5)	(1, 3, 5)
	SD	(1/5, 1/3, 1)	(1/5, 1/3, 1)	(1, 1, 1)	(1/5, 1/3, 1)	(1, 1, 1)	(1/5, 1/3, 1)	(1, 1, 1)	(1, 3, 5)
	SB	(1/5, 1/3, 1)	(1/5, 1/3, 1)	(1, 1, 1)	(1/5, 1/3, 1)	(1, 1, 1)	(1/5, 1/3, 1)	(1/5, 1/3, 1)	(1, 1, 1)

DM$_7$ $\lambda=0.09$	ARW	(1, 1, 1)	(3, 5, 7)	(1, 3, 5)	(1, 3, 5)	(1, 3, 5)	(1, 3, 5)	(1, 3, 5)
	S	(1/7, 1/5, 1/3)	(1, 1, 1)	(1, 3, 5)	(1, 3, 5)	(1, 3, 5)	(1, 3, 5)	(1, 3, 5)
	Max	(1/5, 1/3, 1)	(1/5, 1/3, 1)	(1, 1, 1)	(1/7, 1/5, 1/3)	(1/7, 1/5, 1/3)	(1/5, 1/3, 1)	(1/5, 1/3, 1)
	Min	(1/5, 1/3, 1)	(1/5, 1/3, 1)	(3, 5, 7)	(1, 1, 1)	(1, 3, 5)	(1, 3, 5)	(1, 3, 5)
	IC	(1/5, 1/3, 1)	(1/5, 1/3, 1)	(3, 5, 7)	(1/5, 1/3, 1)	(1, 1, 1)	(1, 1, 1)	(1, 1, 1)
	RL	(1/5, 1/3, 1)	(1/5, 1/3, 1)	(1, 3, 5)	(1, 1, 1)	(3, 5, 7)	(1, 3, 5)	(1, 3, 5)
	SD	(1/5, 1/3, 1)	(1/5, 1/3, 1)	(1, 3, 5)	(1/5, 1/3, 1)	(1, 1, 1)	(1, 1, 1)	(1, 3, 5)
	SB	(1/5, 1/3, 1)	(1/5, 1/3, 1)	(1, 3, 5)	(1/5, 1/3, 1)	(1, 1, 1)	(1, 1, 1)	(1, 1, 1)
DM$_8$ $\lambda=0.08$	ARW	(1, 1, 1)	(1, 1, 1)	(1, 3, 5)	(1, 3, 5)	(1, 3, 5)	(1, 3, 5)	(1, 3, 5)
	S	(1, 1, 1)	(1, 1, 1)	(3, 5, 7)	(1, 3, 5)	(1, 3, 5)	(1, 3, 5)	(1, 3, 5)
	Max	(1/7, 1/5, 1/3)	(1/7, 1/5, 1/3)	(1, 1, 1)	(1/7, 1/5, 1/3)	(1/7, 1/5, 1/3)	(1/7, 1/5, 1/3)	(1/5, 1/3, 1)
	Min	(1/5, 1/3, 1)	(1/5, 1/3, 1)	(1, 3, 5)	(1, 3, 5)	(1, 1, 1)	(1, 3, 5)	(1, 3, 5)
	IC	(1/5, 1/3, 1)	(1/5, 1/3, 1)	(3, 5, 7)	(1, 1, 1)	(1/7, 1/5, 1/3)	(1, 1, 1)	(1, 1, 1)
	RL	(1, 1, 1)	(1, 1, 1)	(3, 5, 7)	(1, 1, 1)	(1, 1, 1)	(1, 3, 5)	(1, 3, 5)
	SD	(1/5, 1/3, 1)	(1/5, 1/3, 1)	(1, 3, 5)	(1/5, 1/3, 1)	(1/5, 1/3, 1)	(1, 1, 1)	(1, 3, 5)
	SB	(1/5, 1/3, 1)	(1/5, 1/3, 1)	(1, 3, 5)	(1/5, 1/3, 1)	(1/5, 1/3, 1)	(1/5, 1/3, 1)	(1, 1, 1)
DM$_9$ $\lambda=0.08$	ARW	(1, 1, 1)	(1, 1, 1)	(1, 3, 5)	(1, 3, 5)	(3, 5, 7)	(1, 3, 5)	(1, 3, 5)
	S	(1, 1, 1)	(1, 1, 1)	(1, 3, 5)	(1, 3, 5)	(1, 3, 5)	(1, 3, 5)	(1, 3, 5)
	Max	(1/5, 1/3, 1)	(1/5, 1/3, 1)	(1, 1, 1)	(1/7, 1/5, 1/3)	(1/5, 1/3, 1)	(1/5, 1/3, 1)	(1/5, 1/3, 1)
	Min	(1/5, 1/3, 1)	(1/5, 1/3, 1)	(3, 5, 7)	(1, 3, 5)	(1, 1, 1)	(1, 3, 5)	(1, 3, 5)
	IC	(1/5, 1/3, 1)	(1/5, 1/3, 1)	(1, 3, 5)	(1, 1, 1)	(1/5, 1/3, 1)	(1, 1, 1)	(1, 1, 1)
	RL	(1/7, 1/5, 1/3)	(1/7, 1/5, 1/3)	(1, 3, 5)	(1, 1, 1)	(1, 1, 1)	(1, 3, 5)	(1, 3, 5)
	SD	(1/5, 1/3, 1)	(1/5, 1/3, 1)	(3, 5, 7)	(1/5, 1/3, 1)	(1/5, 1/3, 1)	(1, 1, 1)	(1, 3, 5)
	SB	(1/5, 1/3, 1)	(1/5, 1/3, 1)	(1, 3, 5)	(1/5, 1/3, 1)	(1/5, 1/3, 1)	(1/5, 1/3, 1)	(1, 1, 1)
DM$_{10}$ $\lambda=0.08$	ARW	(1, 1, 1)	(1, 1, 1)	(1, 3, 5)	(1, 3, 5)	(1, 3, 5)	(1, 3, 5)	(1, 3, 5)
	S	(1, 1, 1)	(1, 1, 1)	(1, 3, 5)	(1, 3, 5)	(1, 3, 5)	(1, 3, 5)	(3, 5, 7)

(Continued)

Table 5.4 (*Continued*) Individual Fuzzy Judgment Matrix for Criteria of Track Selection

		ARW	S	MAX	MIN	IC	RL	SD	SB
	Max	(1/5, 1/3, 1)	(1/5, 1/3, 1)	(1, 1, 1)	(1, 1, 1)	(1, 1, 1)	(1, 1, 1)	(1, 1, 1)	(1, 1, 1)
	Min	(1/5, 1/3, 1)	(1/5, 1/3, 1)	(1, 1, 1)	(1, 1, 1)	(1, 3, 5)	(1, 1, 1)	(1, 3, 5)	(1, 3, 5)
	IC	(1/5, 1/3, 1)	(1/5, 1/3, 1)	(1, 1, 1)	(1/5, 1/3, 1)	(1, 1, 1)	(1, 1, 1)	(1/5, 1/3, 1)	(1/5, 1/3, 1)
	RL	(1/5, 1/3, 1)	(1/5, 1/3, 1)	(1, 1, 1)	(1, 1, 1)	(1, 1, 1)	(1, 1, 1)	(1, 3, 5)	(7, 9, 9)
	SD	(1/5, 1/3, 1)	(1/5, 1/3, 1)	(1, 1, 1)	(1/5, 1/3, 1)	(1, 3, 5)	(1/5, 1/3, 1)	(1, 1, 1)	(1, 3, 5)
	SB	(1/5, 1/3, 1)	(1/7, 1/5, 1/3)	(1, 1, 1)	(1/5, 1/3, 1)	(1, 3, 5)	(1/9, 1/9, 1)	(1/5, 1/3, 1)	(1, 1, 1)
DM$_{11}$ λ = 0.08	ARW	(1, 1, 1)	(1, 1, 1)	(1, 3, 5)	(1, 3, 5)	(1, 3, 5)	(1, 3, 5)	(1, 3, 5)	(1, 3, 5)
	S	(1, 1, 1)	(1, 1, 1)	(1, 3, 5)	(1, 3, 5)	(1, 3, 5)	(1, 1, 1)	(1, 3, 5)	(1, 3, 5)
	Max	(1/5, 1/3, 1)	(1/5, 1/3, 1)	(1, 1, 1)	(1, 1, 1)	(1, 3, 5)	(1, 1, 1)	(1, 3, 5)	(1, 1, 1)
	Min	(1/5, 1/3, 1)	(1/5, 1/3, 1)	(1, 1, 1)	(1, 1, 1)	(1, 3, 5)	(1, 1, 1)	(1, 3, 5)	(1, 3, 5)
	IC	(1/5, 1/3, 1)	(1/5, 1/3, 1)	(1/5, 1/3, 1)	(1/5, 1/3, 1)	(1, 1, 1)	(1, 3, 5)	(1, 3, 5)	(1, 3, 5)
	RL	(1/5, 1/3, 1)	(1, 1, 1)	(1, 1, 1)	(1, 1, 1)	(1, 1, 1)	(1, 1, 1)	(1, 1, 1)	(1, 3, 5)
	SD	(1/5, 1/3, 1)	(1/5, 1/3, 1)	(1/5, 1/3, 1)	(1/5, 1/3, 1)	(1/5, 1/3, 1)	(1/5, 1/3, 1)	(1, 1, 1)	(1, 3, 5)
	SB	(1/5, 1/3, 1)	(1/5, 1/3, 1)	(1, 1, 1)	(1/5, 1/3, 1)	(1/5, 1/3, 1)	(1/5, 1/3, 1)	(1/5, 1/3, 1)	(1, 1, 1)
DM$_{12}$ λ = 0.08	ARW	(1, 1, 1)	(1, 1, 1)	(1, 3, 5)	(1, 3, 5)	(1, 3, 5)	(1, 3, 5)	(1, 3, 5)	(1, 1, 1)
	S	(1, 1, 1)	(1, 1, 1)	(1, 3, 5)	(1, 3, 5)	(1, 3, 5)	(1, 3, 5)	(1, 3, 5)	(1, 1, 1)
	Max	(1/5, 1/3, 1)	(1/5, 1/3, 1)	(1, 1, 1)	(1/7, 1/5, 1/3)	(1/5, 1/3, 1)	(1/5, 1/3, 1)	(1, 1, 1)	(1, 1, 1)
	Min	(1/5, 1/3, 1)	(1/5, 1/3, 1)	(3, 5, 7)	(1, 1, 1)	(1, 3, 5)	(1, 3, 5)	(1, 3, 5)	(1, 3, 5)
	IC	(1/5, 1/3, 1)	(1/5, 1/3, 1)	(1, 3, 5)	(1/5, 1/3, 1)	(1, 1, 1)	(1, 3, 5)	(1, 3, 5)	(1, 3, 5)
	RL	(1/5, 1/3, 1)	(1/5, 1/3, 1)	(1, 3, 5)	(1/5, 1/3, 1)	(1/5, 1/3, 1)	(1, 1, 1)	(1, 3, 5)	(1, 3, 5)
	SD	(1/5, 1/3, 1)	(1/5, 1/3, 1)	(1, 1, 1)	(1/5, 1/3, 1)	(1/5, 1/3, 1)	(1/5, 1/3, 1)	(1, 1, 1)	(1, 3, 5)
	SB	(1, 1, 1)	(1, 1, 1)	(1, 1, 1)	(1/5, 1/3, 1)	(1/5, 1/3, 1)	(1/5, 1/3, 1)	(1/5, 1/3, 1)	(1, 1, 1)

Table 5.5 Individual Fuzzy Priority Vector of Decision Makers and Aggregated Weight Vector for Criteria of Track Selection

	ARW	S	MAX	MIN	IC	RL	SD	SB
DM$_1$	(0.24, 0.25, 0.25)	(0.17, 0.18, 0.16)	(0.06, 0.05, 0.06)	(0.12, 0.11, 0.10)	(0.12, 0.11, 0.12)	(0.18, 0.20, 0.17)	(0.07, 0.06, 0.08)	(0.05, 0.04, 0.06)
DM$_2$	(0.16, 0.20, 0.19)	(0.13, 0.17, 0.19)	(0.11, 0.09, 0.09)	(0.13, 0.14, 0.16)	(0.09, 0.09, 0.10)	(0.16, 0.13, 0.10)	(0.16, 0.12, 0.08)	(0.05, 0.05, 0.08)
DM$_3$	(0.26, 0.32, 0.30)	(0.15, 0.13, 0.12)	(0.15, 0.14, 0.13)	(0.14, 0.14, 0.12)	(0.08, 0.08, 0.10)	(0.10, 0.09, 0.10)	(0.07, 0.06, 0.08)	(0.05, 0.04, 0.06)
DM$_4$	(0.23, 0.28, 0.26)	(0.19, 0.21, 0.21)	(0.13, 0.12, 0.12)	(0.11, 0.10, 0.11)	(0.09, 0.08, 0.09)	(0.13, 0.09, 0.08)	(0.06, 0.07, 0.07)	(0.05, 0.04, 0.06)
DM$_5$	(0.22, 0.30, 0.29)	(0.16, 0.18, 0.18)	(0.13, 0.08, 0.07)	(0.13, 0.12, 0.12)	(0.07, 0.05, 0.07)	(0.15, 0.13, 0.11)	(0.08, 0.07, 0.08)	(0.07, 0.06, 0.08)
DM$_6$	(0.22, 0.29, 0.29)	(0.15, 0.20, 0.20)	(0.13, 0.08, 0.07)	(0.13, 0.12, 0.12)	(0.10, 0.07, 0.07)	(0.13, 0.11, 0.10)	(0.07, 0.08, 0.08)	(0.07, 0.05, 0.07)
DM$_7$	(0.23, 0.29, 0.27)	(0.16, 0.19, 0.18)	(0.05, 0.03, 0.05)	(0.15, 0.14, 0.15)	(0.13, 0.08, 0.08)	(0.14, 0.12, 0.12)	(0.07, 0.08, 0.08)	(0.07, 0.06, 0.08)
DM$_8$	(0.18, 0.21, 0.19)	(0.20, 0.22, 0.20)	(0.04, 0.03, 0.05)	(0.12, 0.14, 0.15)	(0.09, 0.07, 0.08)	(0.23, 0.20, 0.17)	(0.08, 0.07, 0.08)	(0.06, 0.06, 0.08)
DM$_9$	(0.23, 0.25, 0.22)	(0.20, 0.23, 0.21)	(0.04, 0.03, 0.05)	(0.15, 0.14, 0.15)	(0.09, 0.07, 0.08)	(0.23, 0.20, 0.17)	(0.08, 0.07, 0.08)	(0.06, 0.06, 0.08)
DM$_{10}$	(0.19, 0.24, 0.23)	(0.21, 0.25, 0.24)	(0.12, 0.08, 0.07)	(0.12, 0.12, 0.13)	(0.07, 0.05, 0.07)	(0.16, 0.12, 0.11)	(0.07, 0.09, 0.10)	(0.06, 0.05, 0.06)
DM$_{11}$	(0.20, 0.25, 0.22)	(0.20, 0.21, 0.18)	(0.13, 0.11, 0.10)	(0.13, 0.12, 0.12)	(0.09, 0.09, 0.12)	(0.13, 0.11, 0.10)	(0.07, 0.06, 0.08)	(0.06, 0.05, 0.07)
DM$_{12}$	(0.21, 0.21, 0.18)	(0.21, 0.21, 0.18)	(0.07, 0.05, 0.06)	(0.16, 0.17, 0.18)	(0.11, 0.12, 0.14)	(0.09, 0.09, 0.12)	(0.06, 0.07, 0.08)	(0.09, 0.06, 0.06)
Aggregated weight	(0.21, 0.34, 0.32)	(0.18, 0.27, 0.26)	(0.09, 0.10, 0.10)	(0.14, 0.18, 0.19)	(0.09, 0.11, 0.13)	(0.14, 0.17, 0.15)	(0.08, 0.11, 0.11)	(0.06, 0.07, 0.10)

Table 5.6 Aggregated Fuzzy Judgment Matrix for Criteria of Track Selection

	ARW	S	MAX	MIN	IC	RL	SD	SB	MAW
ARW	(1.00, 1.00, 1.00)	(0.22, 2.14, 2.82)	(1.29, 2.85, 4.20)	(1.19, 3.27, 5.31)	(1.00, 3.00, 5.00)	(1.13, 2.31, 3.39)	(1.00, 2.52, 3.88)	(1.00, 2.75, 4.40)	0.27
S	(0.35, 0.47, 0.82)	(1.00, 1.00, 1.00)	(0.14, 2.92, 4.58)	(1.00, 2.75, 4.40)	(1.00, 3.00, 5.00)	(1.11, 1.76, 2.21)	(1.00, 2.52, 3.88)	(1.09, 2.86, 4.52)	0.22
Max	(0.24, 0.35, 0.78)	(0.22, 0.34, 0.88)	(1.00, 1.00, 1.00)	(0.37, 0.50, 0.74)	(0.51, 0.79, 1.19)	(0.54, 0.70, 1.04)	(0.67, 1.04, 1.59)	(0.65, 1.09, 1.66)	0.09
Min	(0.19, 0.31, 0.84)	(0.23, 0.36, 1.00)	(1.35, 1.98, 2.68)	(1.00, 1.00, 1.00)	(1.00, 2.87, 4.69)	(1.00, 1.19, 1.29)	(1.00, 2.52, 3.88)	(1.00, 3.00, 5.00)	0.16
IC	(0.20, 0.33, 1.00)	(0.20, 0.33, 1.00)	(0.84, 1.27, 1.94)	(0.21, 0.35, 1.00)	(1.00, 1.00, 1.00)	(0.61, 0.83, 1.18)	(0.88, 1.37, 1.81)	(0.76, 1.48, 2.33)	0.10
RL	(0.29, 0.43, 0.88)	(0.45, 0.57, 0.90)	(0.96, 1.43, 1.86)	(0.78, 0.84, 1.00)	(0.85, 1.20, 1.64)	(1.00, 1.00, 1.00)	(1.00, 2.52, 3.88)	(1.17, 3.27, 5.24)	0.15
SD	(0.26, 0.40, 1.00)	(0.26, 0.40, 1.00)	(0.45, 1.76, 0.90)	(0.26, 0.40, 1.00)	(0.55, 0.73, 1.14)	(0.26, 0.40, 1.00)	(1.00, 1.00, 1.00)	(1.00, 2.52, 3.88)	0.10
SB	(0.23, 0.36, 1.00)	(0.22, 0.35, 0.92)	(0.60, 0.91, 1.54)	(0.20, 0.33, 1.00)	(0.43, 0.68, 1.32)	(0.19, 0.31, 0.86)	(0.26, 0.40, 1.00)	(1.00, 1.00, 1.00)	0.07

CCI = 0.01.

$S_B = (3.13, 4.34, 8.63) \otimes (1/52.64, 1/100.22, 1/153.81) = (0.02, 0.05, 0.19)$

$V(S_{ARW} \geq S_S) = V(S_{ARW} \geq S_{Max}) = V(S_{ARW} \geq S_{Min}) = V(S_{ARW} \geq S_{IC}) = V(S_{ARW} \geq S_{RL}) = V(S_{ARW} \geq S_{SD}) = V(S_{ARW} \geq S_B) = 1$

$V(S_S \geq S_{ARW}) = V(S_S \geq S_{Min}) = V(S_S \geq S_{IC}) = V(S_S \geq S_{RL}) = V(S_S \geq S_{SD}) = V(S_S \geq S_B) = 1$

$V(S_S \geq S_{Max}) = (0.07 - 0.23)/(0.20 - 0.58) - (0.23 - 0.07) = 0.95$

$V(S_{Max} \geq S_{ARW}) = 0.47, \quad V(S_{Max} \geq S_S) = 0.52, \quad V(S_{Max} \geq S_{Min}) = 0.63, \quad V(S_{Max} \geq S_{IC}) = 0.92$

$V(S_{Max} \geq S_{RL}) = 0.69, V(S_{Max} \geq S_{SD}) = 0.91, V(S_{Max} \geq S_B) = 1$

$V(S_{Min} \geq S_{ARW}) = 0.86, \quad V(S_{Min} \geq S_S) = 0.91, \quad V(S_{Min} \geq S_{Max}) = V(S_{Min} \geq S_{IC}) = V(S_{Min} \geq S_{RL}) = V(S_{Min} \geq S_{SD}) = V(S_S \geq S_B) = 1$

$V(S_{IC} \geq S_{ARW}) = 0.57, \quad V(S_{IC} \geq S_S) = 0.63, \quad V(S_{IC} \geq S_{Max}) = 0.73, \quad V(S_{IC} \geq S_{Min}) = 0.73, \quad V(S_{IC} \geq S_{RL}) = 0.79, \quad V(S_{IC} \geq S_{SD}) = 0.99, \quad V(S_{IC} \geq S_B) = 1$

$V(S_{RL} \geq S_{ARW}) = 0.79, \quad V(S_{RL} \geq S_S) = 0.85, \quad V(S_{RL} \geq S_{Max}) = 0.94, \quad V(S_{RL} \geq S_{Min}) = 0.94, V(S_{RL} \geq S_{IC}) = V(S_{RL} \geq S_{SD}) = V(S_{RL} \geq S_B) = 1$

$V(S_{SD} \geq S_{ARW}) = 0.54, \quad V(S_{SD} \geq S_S) = 0.60, \quad V(S_{SD} \geq S_{Max}) = 0.71, \quad V(S_{SD} \geq S_{Min}) = 0.71, \quad V(S_{SD} \geq S_{IC}) = 1, \quad V(S_{SD} \geq S_{RL}) = 0.77, \quad V(S_{SD} \geq S_B) = 1$

$V(S_B \geq S_{ARW}) = 0.40, \quad V(S_B \geq S_S) = 0.46, \quad V(S_B \geq S_{Max}) = 0.55, \quad V(S_B \geq S_{Min}) = 0.55, \quad V(S_B \geq S_{IC}) = 0.80, \quad V(S_B \geq S_{RL}) = 0.60, \quad V(S_B \geq S_{SD}) = 0.80$

Calculation of the priority weights for criteria is completed by using Equation 5.7:

$d'(ARW) = \min(1, 1, 1, 1, 1, 1, 1) = 1$
$d'(S) = \min(0.95, 1, 1, 1, 1, 1, 1) = 0.95$
$d'(Max) = \min(0.47, 0.52, 0.63, 0.92, 0.69, 0.91, 1) = 0.47$
$d'(Min) = \min(0.86, 0.91, 1, 1, 1, 1, 1) = 0.86$
$d'(IC) = \min(0.57, 0.63, 0.73, 0.73, 0.79, 0.99, 1) = 0.79$
$d'(RL) = \min(0.79, 0.85, 0.94, 0.94, 1, 1, 1) = 0.79$
$d'(SD) = \min(0.54, 0.60, 0.71, 0.71, 1, 0.77, 1) = 0.54$
$d'(B) = \min(0.40, 0.46, 0.55, 0.55, 0.80, 0.60, 0.80) = 0.40$

If it is normalized, the priority weight is computed as $d(C) = (0.18, 0.17, 0.08, 0.15, 0.10, 0.14, 0.10, 0.07)$.

Table 5.7 Aggregated Fuzzy Judgment Matrix for Alternatives of Track Selection under Each Criterion

CRITERIA		T_1	T_2	T_3	MAW
ARW	T_1	(1, 1, 1)	(0.33, 0.60, 1.09)	(0.30, 0.45, 0.70)	0.19
	T_2	(0.91, 1.67, 3.01)	(1, 1, 1)	(0.30, 0.35, 0.47)	0.26
	T_3	(1.43, 2.23, 3.34)	(2.12, 2.86, 3.39)	(1, 1, 1)	0.55
$GCI=0.03$					
Slot	T_1	(1, 1, 1)	(0.53, 0.68, 0.97)	(0.50, 0.71, 1.04)	0.26
	T_2	(1.03, 1.48, 1.90)	(1, 1, 1)	(0.45, 0.55, 0.79)	0.31
	T_3	(0.96, 1.40, 1.98)	(1.27, 1.83, 2.22)	(1, 1, 1)	0.44
$GCI=0.08$					
Max	T_1	(1, 1, 1)	(0.23, 0.36, 075)	(0.30, 0.49, 0.98)	0.19
	T_2	(1.34, 2.81, 4.41)	(1, 1, 1)	(0.32, 0.43, 0.66)	0.32
	T_3	(1.02, 2.06, 3.29)	(1.53, 2.33, 3.15)	(1, 1, 1)	0.49
$GCI=0.06$					
Min	T_1	(1, 1, 1)	(0.39, 0.52, 0.82)	(0.35, 0.44, 0.62)	0.20
	T_2	(1.22, 1.91, 2.59)	(1, 1, 1)	(0.65, 0.79, 0.98)	0.36
	T_3	(1.61, 2.29, 2.84)	(1.03, 1.27, 1.54)	(1, 1, 1)	0.44
$GCI=0.01$					
IC	T_1	(1, 1, 1)	(0.45, 0.64, 1.06)	(0.39, 0.51, 0.76)	0.22
	T_2	(0.94, 1.57, 2.23)	(1, 1, 1)	(0.67, 0.99, 1.24)	0.36
	T_3	(1.32, 1.98, 2.58)	(0.81, 1.01, 1.50)	(1, 1, 1)	0.41
$GCI=0.05$					
RL	T_1	(1, 1, 1)	(0.36, 0.57, 0.99)	(0.37, 0.50, 0.75)	0.21
	T_2	(1.01, 1.75, 2.76)	(1, 1, 1)	(0.37, 0.51, 0.74)	0.30
	T_3	(1.33, 2.01, 2.70)	(1.34, 1.98, 2.74)	(1, 1, 1)	0.49
$GCI=0.03$					
SD	T_1	(1, 1, 1)	(0.41, 0.61, 1.04)	(0.39, 0.51, 0.77)	0.23
	T_2	(0.97, 1.64, 2.43)	(1, 1, 1)	(0.59, 0.92, 1.28)	0.36
	T_3	(1.30, 1.95, 2.59)	(0.78, 1.09, 1.68)	(1, 1, 1)	0.41
$GCI=0.04$					
SB	T_1	(1, 1, 1)	(0.36, 0.57, 0.99)	(0.36, 0.58, 0.95)	0.22
	T_2	(1.01, 1.75, 2.76)	(1, 1, 1)	(0.32, 0.47, 0.71)	0.29
	T_3	(1.05, 1.73, 2.81)	(1.41, 2.12, 3.11)	(1, 1, 1)	0.49
$GCI=0.03$					

Then, similar steps are followed for the alternatives. Table 5.7 indicates the aggregated fuzzy judgment matrix under each criterion, which is calculated from the DMs' individual fuzzy judgment matrices.

Table 5.8 presents the final outputs of the route selection problem. Track 3 is found to be the most feasible route by the AHP expert

Table 5.8   Final Assessment of Alternatives of Track Selection

	ARW	S	MAX	MIN	IC	RL	SD	SB	ALTERNATIVE PRIORITY
Weight	0.18	0.17	0.08	0.15	0.10	0.14	0.10	0.07	Weight
T_1	0.05	0.19	0.14	0.00	0.17	0.12	0.18	0.18	0.12
T_2	0.30	0.31	0.39	0.43	0.39	0.35	0.39	0.34	0.35
T_3	0.65	0.50	0.46	0.57	0.44	0.53	0.44	0.49	0.52[a]

[a] Selected alternative.

consultation. Superiority of the selected route is quite explicit as the difference between the first and second selection is 0.17.

5.5 Conclusion

In the traditional approach, the shortest sea route is usually preferred as the cost aversion drives DMs, especially for ice navigation because less time can be spent in ice. On the other hand, the safety of the route is a subjective factor that cannot be directly measured and evaluated. By using the F-AHP method, the safety risk is indirectly embedded into the decision-making process by consulting with experts. Navigational safety in the Arctic region is mostly related with the dimensional limitations of route. The empirical results exposed an opposite ranking rather than the traditional expectations. The shortest sea route (track 1) is the last optimum while the longest route (track 3) is the best among three alternatives.

It is clear that the shortest navigational route does not guarantee the safety of navigation and the group of experts in the field also agreed on the objective of this study by defining a difference between the length of the route and other dimensions.

References

Aguarón, J. and Moreno-Jiménez, J.M. 2003. The geometric consistency index: Approximated thresholds. *European Journal of Operational Research* 147:137–145.

Ari, I., Aksakalli, V., Aydogdu, V., and Kum, S. 2013. Optimal ship navigation with safety distance and realistic turn constraints. *European Journal of Operational Research* 229:707–717.

Bellman, R.E. and Zadeh, L.A. 1977. *Local and Fuzzy Logics. Modern Uses of Multiple-Valued Logic.* Boston, MA: Kluwer Academic Publishers B.V., pp. 105–151.

Buckley, J.J. 1985. Fuzzy hierarchical analysis. *Fuzzy Sets and System* 17:233–247.

Bulut, E., Duru, O., Kececi, T., and Yoshida, S. 2012. Use of consistency index, expert prioritization and direct numerical inputs for generic fuzzy-AHP modelling: A process model for shipping asset management. *Expert Systems with Applications* 39:1911–1923.

Buysee, J. 2007. *Handling Ships in Ice: A Practical Guide to Handling Class 1A and 1AS Ships.* London, U.K.: The Nautical Institute Press.

Cakir, O. and Canpolat, M.S. 2008. A web-based decision support system for multi-criteria inventory classification using fuzzy AHP methodology. *Expert Systems with Applications* 35:1367–1378.

Chan, F.T.S., Chan, M.H., and Tamg, N.K.H. 2000. Evaluation methodologies for technology selection. *Journal of Materials Processing Technology* 107:330–337.

Chang, D.Y. 1996. Applications of the extent analysis method on fuzzy AHP. *European Journal of Operational Research* 95:649–655.

Chen, C.T. 2000. Extensions of the TOPSIS for group decision making under fuzzy environment. *Fuzzy Sets and Systems* 114:1–9.

Chiclana, F., Herrera, F., and Herrera-Viedma, E. 1998. Integrating three representation models in fuzzy multipurpose decision making based on fuzzy preference relations. *Fuzzy Sets and Systems* 97:33–48.

Crawford, G. and Williams, C. 1985. A note on the analysis of subjective judgment matrices. *Journal of Mathematical Psychology* 29:387–405.

Duru, O., Bulut, E., and Yoshida, S. 2012. Regime switching fuzzy AHP model for choice-varying priorities problem and expert consistency prioritization: A cubic fuzzy-priority matrix design. *Expert Systems with Applications* 39:4954–4964.

Gumus, A.T. 2009. Evaluation of hazardous waste transportation firms by using a two step fuzzy-AHP and TOPSIS methodology. *Expert Systems with Applications* 36:4067–4074.

Kahraman, C., Cebeci, U., and Ruan, D. 2004. Multi-attribute comparison of catering service companies using fuzzy AHP: The case of Turkey. *International Journal of Production Economics* 87:171–184.

Laarhoven, P.J.M. and Pedrycz, W. 1983. A fuzzy extension of Saaty's priority theory. *Fuzzy Sets and Systems* 11:229–241.

Lee, M., Pham, H., and Zhang, X. 1999. A methodology for priority setting with application to software development process. *European Journal of Operational Research* 118:375–389.

Leung, L.C. and Cao, D. 2000. On consistency and ranking of alternatives in fuzzy AHP. *European Journal of Operational Research* 124:102–113

Saaty, T.L. 1977. A scaling method for priorities in hierarchical structures. *Journal of Mathematical Psychology* 15:234–281.

Saaty, T.L. 1980. *Multi-Criteria Decision Making: The Analytic Hierarchy Process.* New York: McGraw-Hill.

Saaty, T.L. and Vargas, L.G. 1987. Uncertainty and rank order in the analytic hierarchy process. *European Journal of Operational Research* 32:107–117.

Schøyen, H. and S. Bråthen. 2011. The Northern Sea Route versus the Suez Canal: Cases from bulk shipping. *Journal of Transport Geography* 19: 977–983.

Thomson, N.R. and Sykes, J.F. 1988. Route selection through a dynamic ice field using the maximum principle. *Transportation Research Part B: Methodological* 22:339–356.

Weck, M., Klocke, F., Schell, H., and Ruenauver, E. 1997. Evaluating alternative production cycles using the extended fuzzy AHP method. *European Journal of Operational Research* 100:351–366.

Wilson, K.J., Falkingham, J., Melling, H., and Abreu, R.D. 2004. Shipping in the Canadian Arctic: other possible climate change scenarios. *Proceedings of the International Geoscience and Remote Sensing Symposium*, Anchorage, AK.

Zadeh, L.A. 1965. Fuzzy and sets. *Information and Control* 8:338–353.

Zadeh, L.A. 1975. The concept of linguistic variable and its application to approximate reasoning. *Information Sciences* 8:199–249.

6

ROUTE DESIGN IN A PHARMACEUTICAL WAREHOUSE VIA MATHEMATICAL PROGRAMMING

ZEYNEL SIRMA, AYSUN AKIŞ, ZEYNEP YALÇIN, AND FADİME ÜNEY-YÜKSEKTEPE

Contents

6.1 Introduction

İstanbul Ecza Deposu was established in 1989 as a logistics company that makes daily medicine distribution to pharmacies. Company has 4000 contractual pharmacies and 14 warehouses, which have their own serving regions. In this study, central warehouse located in Bahçelievler, Istanbul, is addressed. Currently, it has 22 predetermined routes serving to contractual pharmacies. While a new pharmacy is added to the system, its route assignment is done according to its physical location and the capacity of each route.

When the number of pharmacies increases, the timing of the orders may be problematic and the customers could be dissatisfied if the route assignments are done inefficiently. The profitability of the specified routes depends on the demands of each pharmacy. In addition, the variability in the specified delivery time negatively affects the firm's customers. Hence, the route design is an important factor that affects both profitability and customer experience. In this study, the problem is to efficiently determine the route assignment for each pharmacy while maximizing the profitability and minimizing the lateness in the delivery time. Therefore, it is a type of vehicle routing problem (VRP) without any capacity restrictions and distinct vehicle types.

Supply chain is a network of facilities and distribution alternatives that manage the functionality of supplement of materials, the planning issues related to production, and the distribution of finished goods to the customer (Cooper et al., 1997). In supply chain, the logistics cost constitutes considerable part of the total supply chain costs. Therefore, efficiently planning the logistics issues will improve the overall performance of a supply chain.

The VRP is the well-known logistics planning problem that tries to find the optimal distribution routes for available vehicles. The main components of VRP are drivers, depots, customers, and vehicles. It is a combinatorial optimization problem widely studied in literature. Eksioglu et al. (2009) gave a taxonomic review of VRPs.

One of the bus network design problems was dealt by Szeto and Wu (2011). Their study's main objective is to develop the current bus services by lowering the transfers and total travel time of the passengers. In solution methodology, a genetic programming heuristic procedure had been used to overcome frequency setting and route design problem. Another article that investigated a VRP model is recently published by Liu et al. (2013). The intelligent-van approach to Telematics system is the problem considered in that study. This system is responsible for distributing pharmaceutical materials such as drugs, from their depot to specified pharmacies through delivery routes without predetermining the assigned pharmacies.

In VRP models, first obtaining clusters of customers and then optimizing the routes based on that clusters is one of the in-use solution methods. Cluster analysis distinguishes a heterogeneous group of records into more homogenous classes (Nispet et al., 2009). The aim of the clustering is to find the objects that are similar to one another and different from the objects in other groups. The major similarities with a group and the major contradistinction between groups allow better or more distinct clustering (Kumar et al., 2006). Özdamar and Demir (2012) described a hierarchical clustering and a routing method (HOGCR) for large-scale disaster relief logistics planning. They used a multilevel clustering algorithm in which demand nodes are divided into compact clusters in all planning steps. After clustering, the routing problem was defined by network flow models and solved on a platform that makes parallel computing. In another study, a capacitated location-routing problem is solved by a greedy clustering method (Mehrjerdi and Nadizadeh, 2013). As customers have fuzzy demands, a fuzzy chance-constrained programming model is proposed to solve the mentioned problem. The efficiency of the new approach is tested on a number of numerical experiments. Yücenur and Demirel (2011) offered a new geometric shape-based genetic clustering algorithm in order to solve a multidepot VRP model. When the performance of the proposed approach is compared with the nearest neighbor algorithm, the model is better in terms of total distances and computational times.

In this study, two mathematical models are established to plan a pharmaceutical company's medicine distribution. First, a mathematical model is developed for clustering the pharmacies according to location similarity, and another model is developed to optimize the transportation of drugs from a pharmaceutical warehouse to the contractual pharmacies depending on their daily demands. By using the data obtained from the company, the routes are reorganized while satisfying restrictions and optimizing its objectives. The developed model is solved by using GAMS software (Brooke et al., 1998) and CPLEX solver (ILOG, 2012). The results are compared with the current situation, and different scenario analyses are performed.

6.2 Problem Definition

Currently, Ecza Koop provides services from 1 head office and 13 branches, which are located in Istanbul. In this project, head office's distribution problem is studied. Company's head office is currently using 22 basic routes to provide drugs to their pharmacies in Istanbul. There are a total of 576 pharmacies in this region. The company has 19 vehicles. Orders are given to company by pharmacies according to the quantity determined the daily periodical agreements. Orders are transferred by drivers usually in the morning, and determinations of the routes are based on drivers' decisions. Moreover, when the number of pharmacies increases, they are included in the currents routes, and the route cycle time is affected. Therefore, the planning and routing become problematic and should be optimized.

6.3 Proposed Mathematical Models

This study covers two mathematical models for clustering and vehicle routing.

6.3.1 Mathematical Model for Clustering

Cluster analysis is a comprehensive concept that is used to obtain groups of pharmacies that are close to each other. Clustering model is formulated to obtain groups of pharmacies that are close to each other. There are two index set in the clustering model:

i,j = pharmacies $(i,j = 1,...,I)$
k = clusters $(k = 1,...,K)$

The following parameters are used in the proposed clustering model:

e_i: latitude of pharmacy i
b_i: longtitude of pharmacy i
p_i: profit of pharmacy i (TL)
P_{max}: maximum profit for each cluster (TL)
dis_{ij}: distance between pharmacy i and pharmacy j (km)

The distance between pharmacy i and pharmacy j is calculated by using the following equation:

$$dis_{ij} = \sqrt{\left(e_i - e_j\right)^2 + \left(b_i - b_j\right)^2} \qquad (6.1)$$

The model's decision variables are listed as follows:

D_{max} = maximum diameter of constructed clusters (km)
D_k = diameter of cluster k (km)

$$X_{ik} = \begin{cases} 1, & \text{if pharmacy } i \text{ is in cluster } k \\ 0, & \text{otherwise} \end{cases}$$

N_{max} = maximum number of pharmacies in a cluster
N_k = number of pharmacies in cluster k
TP_k = total profit of cluster k

The following mathematical model is developed to obtain the clusters of pharmacies:

$$\min z = D_{max} + N_{max} \tag{6.2}$$

subject to

$$D_k \geq dis_{ij} \cdot \left(X_{ik} + X_{jk} - 1 \right) \quad \forall i, j, k \tag{6.3}$$

$$\sum_k X_{ik} = 1 \quad \forall i \tag{6.4}$$

$$D_k \leq D_{max} \quad \forall k \tag{6.5}$$

$$\sum_i X_{ik} = N_k \quad \forall k \tag{6.6}$$

$$N_k \leq N_{max} \quad \forall k \tag{6.7}$$

$$TP_k = \sum_i p_i \cdot X_{ik} \quad \forall k \tag{6.8}$$

$$TP_k \leq P_{max} \quad \forall k \tag{6.9}$$

$$D_{max}, N_{max} \geq 0 \tag{6.10}$$

$$D_k, TP_k, N_k \geq 0 \quad \forall k \tag{6.11}$$

$$X_{ik} \in \{0,1\} \quad \forall i, k \tag{6.12}$$

In this model, objective function given in Equation 6.2 is defined as minimizing the total of the maximum diameter of constructed clusters and the maximum number of pharmacies in each cluster. Constraint (6.3) ensures that if two pharmacies j and k are in the same cluster, the distance between them should be lower than the maximum diameter of that cluster. Constraint (6.4) shows that each pharmacy should be assigned to a single cluster. The maximum diameter of overall clusters is calculated by using Equation 6.5. Equation 6.6 calculates the total number of pharmacies in each cluster. The maximum number of pharmacies in constructed clusters is obtained by using Equation 6.7. Equation 6.8 calculates the profit of each one of the clusters. Equation 6.9 gives the maximum profit restriction to each one of the clusters. Equations 6.10 through 6.12 give the integrality and nonnegativity of the decision variables.

6.3.2 Mathematical Model for Vehicle Routing

The VRP defines routes for available vehicles. The main components of VRP are drivers, customers, and vehicles. The main purpose of using VRP in this study is to find the optimal routes for distribution while maximizing total profit. The following indices and parameters are used in the proposed mathematical model.

i,j: pharmacies $(i,j = 1,...,I)$
d_{ij}: distance from node i to node j (m)
p_i: monthly average profit of node i (TL)
fcw: monthly fixed cost of workers (TL)
fcv: monthly fixed cost of vehicles (TL)
vcv: variable cost of vehicles (TL/m)
ttl: travel time limit for each route (min)
nt: number of tours for each vehicle per day
nd: number of days serviced in a month
nr: total number of routes
s: average speed of vehicles (m/min)
M: a large number
$h_{ij} = \begin{cases} 1, & \text{if pharmacy } i \text{ and pharmacy } j \text{ are in the same cluster} \\ 0, & \text{otherwise} \end{cases}$

The model's decision variables are listed as follows:

$$X_{ij} = \begin{cases} 1, & \text{if pharmacy } j \text{ is visited after pharmacy } i \\ 0, & \text{otherwise} \end{cases}$$

A_i = arrival time of vehicle to pharmacy i (min)

The model of the VRP is as follows:

$$\max z = \sum_i p_i - nr * (fcw + fcv) - \sum_i \sum_j X_{ij} * d_{ij} * vcv * nt * nd$$

$$\tag{6.13}$$

$$\sum_i h_{ij} * X_{ij} = 1 \quad \forall j : j \neq 1, j \neq I \tag{6.14}$$

$$\sum_j h_{ij} * X_{ij} = 1 \quad \forall i : i \neq 1, i \neq I \tag{6.15}$$

$$\sum_i h_{i,I} * X_{i,I} = nr \tag{6.16}$$

$$\sum_j h_{1,j} * X_{1,j} = nr \tag{6.17}$$

$$A_1 = 0 \tag{6.18}$$

$$A_j \geq A_i + \frac{d_{ij}}{s} - M(1 - X_{ij}) \quad \forall i, j \tag{6.19}$$

$$A_i + \frac{d_{i,I}}{s} \leq ttl \quad \forall i : i \neq I \tag{6.20}$$

$$A_i \geq 0 \quad \forall i \tag{6.21}$$

$$X_{ij} \in \{0,1\} \quad \forall i, j \tag{6.22}$$

In this model, the objective function given in Equation 6.13 is to maximize the total profit. Equation 6.14 denotes that for each pharmacy, there must be exactly one incoming arc which originates from the warehouse or another pharmacies. Equation 6.15 denotes that there must be exactly one outgoing arc for each pharmacy, which goes to other pharmacies or the warehouse.

Equations 6.16 and 6.17 ensure that the total number of incoming and outgoing arcs to and from the warehouse should be equal to the total number of roads, respectively. Equation 6.18 shows that the arrival time of the warehouse is 0. Equation 6.19 is used to calculate the arrival time of each location. Equation 6.20 ensures that the total traveling time of each vehicle has to be equal or less than the maximum duration of a vehicle trip. Equations 6.21 and 6.22 give the integrality and nonnegativity of decision variables.

6.4 Computational Results

This study is implemented by using the real data obtained from İstanbul Ecza Koop. Among the current routes, three neighbor routes of them are problematic. Therefore, the pharmacies that exist in those routes are considered in the models. There are 71 pharmacies that should be clustered, and new sequence of visits should be determined. The necessary coordinates of the pharmacies and distance calculations are performed by the help of Google Earth software.

6.4.1 Results of Clustering Model

The developed mathematical models are solved by GAMS software (Brooke et al., 1998) and CPLEX solver (ILOG, 2012). The optimal solution of the models gives that there are 23 pharmacies in Cluster 1, 24 pharmacies in Cluster 2, and 24 pharmacies in Cluster 3. The diameter, number of pharmacies, and profit for each cluster are shown in Table 6.1.

Table 6.1 Diameter, Number of Pharmacies, and Profit for Each Cluster

ITEMS	CLUSTER 1	CLUSTER 2	CLUSTER 3
Number of pharmacies	23	24	24
Diameter (km)	0.099	0.082	0.084
Profit (TL)	251,216	370,943	369,531

6.4.2 Results of VRP

The main purpose of using VRP in this study is to find the optimal routes for distribution while maximizing total profit. The following assumptions are considered for VRP model:

- The company has single vehicle dedicated to each route.
- Monthly average profits for the year 2012 are used.
- The number of tours per day of each vehicle is assumed as 4.
- The total number of days serviced in a month is 22.
- Average speed of vehicles is assumed as 50 km/h.
- For each tour, maximum time limit is 45 min.
- Fixed costs of vehicles and drivers are taken as 933 and 1800 TL, respectively.
- Service time is negligible.

The proposed VRP model is solved by using GAMS software (Brooke et al., 1998) and CPLEX 12.0 solver (ILOG, 2012), and the optimal routes are determined. The proposed mathematical model is solved for three different scenarios. At each scenario, different traveling time limit is used for the constructed routes. Table 6.2 gives the result of these scenarios and the comparison of the current and proposed solutions. For each scenario, traveling time for each cluster is lower than the current solution's traveling time values. Moreover, the total profit is better than the current solution as the visiting sequences of the pharmacies are optimized. Therefore, using the proposed mathematical models will improve the drug distribution system and increase profits.

Furthermore, the characteristics of the proposed mathematical models are given in Table 6.3. As it is seen, the computational times are considerably low, and the optimal solutions are obtained in a very short amount of time.

Table 6.2 Comparison of Current and Proposed Solutions

COMPARISONS	PROPOSED SOLUTION			CURRENT PLAN		
	ROUTE 1	ROUTE 2	ROUTE 3	ROUTE 1	ROUTE 2	ROUTE 3
Number of pharmacies	24	24	23	24	24	23
Traveling time (limit: 40 min)	33.71	28.74	26.86	80	90	90
Traveling time (limit: 50 min)	40.45	34.49	32.23	80	90	90
Traveling time (limit: 60 min)	50.56	43.11	40.29	80	90	90
Total profit (TL)	958,813			938,778		

Table 6.3 Characteristics of the Proposed Mathematical Models

ITEMS	CLUSTER MODEL	VRP MODEL
Number of binary variables	213	5547
Number of continuous variables	12	5329
Number of constraints	15,210	74
Number of iterations	10,827	1986
Computational time (CPUs)	0.421	0.717

6.5 Conclusion

The logistic planning problem of a pharmaceutical warehouse is studied as an optimization problem. The problem is taken from İstanbul Ecza Koop, a pharmaceutical warehouse that distributes medicines to the contracted pharmacies. As the number of pharmacies increases, the timing of the orders may be problematic, and the customers could be dissatisfied if the route assignments are done inefficiently. The profitability of the specified routes depends on the demands of each pharmacy and the distribution-related costs. In addition, the variability in the specified delivery time negatively affects the firm's customers. Hence, the route design is an important factor that affects both profitability and customer experience.

Therefore, in order to find an optimal solution for this problem, two mathematical models are proposed: clustering and vehicle routing models. The main objective of the clustering model is to minimize the total of the maximum diameter of constructed clusters and the maximum number of pharmacies in each cluster. In clustering model, coordinates of pharmacies are used as an input data. As a result, three clusters are determined with approximately equal number of pharmacies in each cluster.

According to the results of clustering, VRP model defines routes for available vehicles. While maximizing the profits of the firm, VRP model is constructed by using company's required constraints and suitable variables. When the results are observed, the efficiency of the proposed solution is observed by comparing the profits.

As a result, pharmaceutical warehouse should optimize their distribution routes in order to decrease their transportation costs and increase their profits. This study shows the positive effect of a systematic approach on a real-life problem of a logistics firm.

References

Brooke, A., Kendric, D., Meraus, A., and Ramon, R. 1998. *GAMS: A User's Guide*. GAMS Development Co., Washington, DC.

Cooper, M.C., Lambert, D.M., and Pagh, J. 1997. Supply chain management, more than a new name for logistics. *The International Journal of Logistics Management* 8 (1): 1–14.

Eksioglu, B., Vural, A.V., and Reisman, A. 2009. The vehicle routing problem: A taxonomic review. *Computers & Industrial Engineering* 57 (4): 1472–1483.

ILOG. 2012. *CPLEX 12.0 User's Manual*. ILOG S.A, New York.

Kumar, V., Steinbach, M., and Tan, P. 2006. *Introduction to Data Mining*. Addison Wesley, Boston, MA.

Liu, S.Y., Cheng, L., Feng, Y.P., and Rong, G. 2013. Clustering structure and logistics: A new framework for supply network analysis. *Chemical Engineering Research and Design* 91 (8): 1383–1389.

Mehrjerdi, Y.Z. and Nadizadeh, A. 2013. Using greedy clustering method to solve capacitated location-routing problem with fuzzy demands. *European Journal of Operational Research* 229: 75–84.

Nispet, R., Elder, J., and Miner, G. 2009. *Handbook of Statistical Analysis and Data Mining Applications*. Elsevier, Oxford, UK.

Özdamar, L. and Demir, O. 2012. A hierarchical clustering and routing procedure for large scale. *Transportation Research Part E* 48: 591–602.

Szeto, W. and Wu, Y. 2011. A simultaneous bus route design and frequency setting problem for Tin Shui Wai, Hong Kong. *European Journal of Operational Research* 209: 141–155.

Yücenur, G.N. and Demirel, N.Ç. 2011. A new geometric shape-based genetic clustering algorithm for the multidepot vehicle routing problem. *Expert Systems with Applications* 38: 11859–11865.

7

INTEGRATED DECISION MODEL FOR MEDICAL SUPPLIER EVALUATION

MEHTAP DURSUN, ZEYNEP SENER, AND E. ERTUGRUL KARSAK

Contents

7.1 Introduction

Supply chain is composed of a complex sequence of processing stages, ranging from raw material supplies, parts manufacturing, components, and end products assembling to the delivery of end products. In the context of supply chain management, supplier selection decision is considered as one of the key issues faced by operations and purchasing managers to remain competitive. A well-selected set of suppliers make a strategic difference to an organization's ability to reduce costs and improve the quality of its end products. Supplier selection and management can be applied to a variety of suppliers throughout a product's life cycle from initial raw material acquisition to end-of-life service providers. Thus, the breadth and diversity of suppliers make the process even more cumbersome (Bai and Sarkis, 2010).

Supplier selection is a strategic decision process that determines the long viability of a company, especially when purchasing costs constitute a significant portion of the operating costs (Hammami et al., 2014). According to the vast literature on supplier selection, the following properties need to be considered while resolving the supplier selection problem (Chen et al., 2006). First, the supplier selection process requires considering multiple conflicting criteria. Second, several decision makers are oftentimes involved in the decision process. Third, decision making is often influenced by uncertainty in practice. Thus, supplier selection that requires considering multiple conflicting criteria incorporating vagueness and imprecision with the involvement of a group of experts is an important multicriteria group decision-making problem. The fuzzy set theory is a viable decision aid that enables to account for the inherent imprecision and vagueness in criteria values. In the literature, there are a number of studies that use various fuzzy decision-making techniques to evaluate suppliers. Several authors have used fuzzy mathematical programming approaches (Kumar et al., 2006; Wu et al., 2010; Ahmady et al., 2013; Nazari-Shirkouhi et al., 2013). A number of studies have focused on the use of fuzzy multiattribute decision-making techniques for supplier selection process (Bottani and Rizzi, 2005; Wang, 2010; Shemshadi et al., 2011; Shen et al., 2013). Lately, a few researchers have employed quality function deployment (QFD) in supplier selection (Bevilacqua et al., 2006; Amin and Razmi, 2009; Alinezad et al., 2013; Dursun and Karsak, 2013).

The objective of this study is to propose a fuzzy multicriteria group decision-making methodology integrating 2-tuple fuzzy linguistic representation model, decision-making trial and evaluation laboratory (DEMATEL) method, and QFD. A house of quality (HOQ) matrix, which translates purchased product features into supplier assessment criteria, is built using the weights obtained by the DEMATEL approach to determine the desired levels of supplier assessment criteria. Finally, supplier alternatives are ranked by a distance-based method.

The rest of this chapter is organized as follows: the following section presents the basic concepts of QFD. In Section 7.3, the DEMATEL method is briefly introduced. Sections 7.4 and 7.5 delineate the fusion of fuzzy information approach and 2-tuple fuzzy linguistic representation model, respectively. Section 7.6 presents the developed decision-making approach and provides its stepwise representation.

The implementation of the proposed framework for evaluating medical suppliers of a private hospital in Istanbul is provided in Section 7.7. Concluding remarks are given in the last section.

7.2 Quality Function Deployment

QFD is a customer-oriented design tool for developing new products to increase customer satisfaction. QFD is also a tool for analyzing and improving manufacturing systems. The basic concept of QFD is to translate the desires of customers into design requirements, and subsequently into parts characteristics, process plans, and production requirements (Karsak, 2004). In order to establish these relationships, QFD usually requires four matrices, each corresponding to a stage of the product development cycle, namely, product planning, part deployment, process planning, and production/operation planning matrices. The product planning matrix translates customer needs (CNs) into technical attributes (TAs), the part deployment matrix translates important TAs into product/part characteristics, the process planning matrix translates important product/part characteristics into manufacturing operations, and the production/operation planning matrix translates important manufacturing operations into day-to-day operations and controls (Shillito, 1994).

The first of the four matrices, also called the HOQ, is the most recognized and widely used matrix in QFD. It translates customer requirements, based on marketing research and benchmarking data, into an appropriate number of engineering targets. Basically, it is the nerve center and the engine that drives the entire QFD process. Relationships between CNs and TAs and among the TAs are defined by answering a specific question corresponding to each cell in HOQ.

The elements of the HOQ can be briefly described as follows:

1. *CNs*: They are also known as voice of the customer, customer attributes, customer requirements, or demanded quality. The initial steps in constructing a HOQ include specifying the CNs. As the initial input for the HOQ, they highlight the product characteristics that should be paid attention to. The CNs can include the requirements of retailers or the needs of vendors.

2. *TAs*: TAs are also named as design requirements, product features, engineering attributes, engineering characteristics, or substitute quality characteristics. They are the product requirements that relate directly to the customer requirements. TAs describe the product in the language of the engineer; therefore, they are sometimes referred to as the voice of the company. They are used to determine how well the company satisfies the CNs (Karsak et al., 2003).

3. *Importance of CNs*: Since the collected and organized data from the customers usually contain too many needs to deal with simultaneously, they must be rated. The company should trade off one benefit against another and work on the most important needs while eliminating relatively unimportant ones (Karsak et al., 2003).

4. *Relationships between CNs and TAs*: The relationship matrix indicates to what extent each TA affects each CN and is placed in the body of the HOQ (Alptekin and Karsak, 2011). In this chapter, linguistic variables are used to denote the relationships between CNs and TAs.

5. *Competitive assessment matrix*: Understanding how customers rate the competition can be a tremendous competitive advantage. The information needed can be obtained by asking the customers to rate the performance of the company's and its competitors' products for each CN using a predetermined scale.

6. *Inner dependence among the TAs*: The HOQ's roof matrix is used to specify the inner dependencies among TAs. This enables to account for the correlations between TAs, which in turn facilitates informed trade-offs.

7. *Overall priorities of the TAs and additional goals*: Here, the results obtained from preceding steps are used to calculate a final rank order of TAs.

7.3 DEMATEL Method

The DEMATEL method was intended to study and resolve the complicated and intertwined problem group. This method could improve understanding of the specific issue, the cluster of intertwined problems, and contribute to the identification of workable solutions by a hierarchical structure. Four major steps of the DEMATEL method can be summarized as follows (Tzeng et al., 2010):

Step 1: Compute the average matrix.

Respondents are asked to indicate the direct influence that they believe each factor i exerts on each factor j of the others, as indicated by a_{ij}. From any group of direct matrices of respondents, it is possible to derive an average matrix A. The diagonal elements of the average matrix are all set to zero, which means no influence is given by itself.

Step 2: Calculate the normalized initial direct-relation matrix.

The normalized initial direct-relation matrix D can be obtained as $D = \xi A$, where

$$\xi = \min \left[\frac{1}{\max_{1 \le i \le n} \sum_{j=1}^{n} |a_{ij}|}, \frac{1}{\max_{1 \le i \le n} \sum_{i=1}^{n} |a_{ij}|} \right] \quad (7.1)$$

Step 3: Calculate the total relation matrix.

The total relation matrix T is defined as $T = D(I-D)^{-1}$, where I is the identity matrix. Define f and c as $n \times 1$ and $1 \times n$ vectors representing the sum of rows and sum of columns of the total relation matrix T, respectively. Suppose f_i be the sum of ith row in matrix T, then f_i summarizes both direct and indirect effects given by factor i to the other factors. If c_j denotes the sum of jth column in matrix T, then c_j shows both direct and indirect effects by factor j from the other factors. When $j = i$, the sum $(f_i + c_j)$ shows the total effects given and received by factor i. Thus, $(f_i + c_j)$ indicates the degree of importance for factor i in the entire system. On the contrary, the difference $(f_i - c_j)$ represents the net effect that factor i contributes to the system. Specifically, if $(f_i - c_j)$ is positive, factor i is a net cause, whereas factor i is a net receiver or result if $(f_i - c_j)$ is negative.

Step 4: Set up a threshold value to obtain the digraph.

In order to explain the structural relation among the factors while keeping the complexity of a system to a manageable level, it is necessary to set a threshold value to filter out some negligible effect in the total relation matrix.

7.4 Fusion of Fuzzy Information

Fusion approach of fuzzy information is proposed by Herrera et al. (2000) to carry out the aggregation step of a decision process in a group decision-making problem defined using nonhomogeneous information.

This approach consists of obtaining a collective performance profile on the alternatives according to the individual performance profiles. It is performed in two phases (Herrera et al., 2000):

1. Making the information uniform
2. Aggregating individual preference values

7.4.1 Making the Information Uniform

The nonhomogeneous information will be unified into a specific linguistic domain, called basic linguistic term set (BLTS) denoted as S_T, chosen so as not to impose useless precision to the original evaluations and to allow an appropriate discrimination of the initial performance values. The process of unifying the information involves the comparison between fuzzy sets. These comparisons are usually carried out by means of a measure of comparison.

The transformation function is defined as follows (Herrera et al., 2000).

Let $\Omega = \{l_0, l_1, \ldots, l_H\}$ and $S_T = \{s_0, s_1, \ldots, s_G\}$ be two linguistic term sets, such that $G \geq H$. Then, the transformation function, τ_{AS_T}, is defined as

$$\tau_{AS_T} : \Omega \rightarrow F(S_T)$$

$$\tau_{AS_T}(l_b) = \left\{ \frac{\left(s_g, \gamma_g^b\right)}{g} \in \{0, 1, \ldots, G\} \right\}, \quad \forall l_b \in \Omega \qquad (7.2)$$

$$\gamma_g^b = \max_y \min\left\{\mu_{l_b}(y), \mu_{s_g}(y)\right\}$$

where
$F(S_T)$ is the set of fuzzy sets defined in S_T
$\mu_{l_b}(y)$ and $\mu_{s_g}(y)$ are the membership functions of the fuzzy sets associated with the terms l_b and s_g, respectively

The transformation function is also appropriate to convert the standardized fuzzy assessments into a BLTS (Chuu, 2009). The max–min operation has been chosen in the definition of the transformation function since it is a classical tool to set the matching degree between fuzzy sets (Herrera et al., 2000).

7.4.2 Aggregating Individual Preference Values

The input information, which was denoted by means of fuzzy sets, is expressed on a BLTS by the earlier-mentioned transformation function. Then, in order to obtain a collective preference value for each alternative, an aggregation function is used. This collective performance value is a new fuzzy set defined on a BLTS.

This chapter employs ordered weighted averaging (OWA) operator, initially proposed by Yager (1988), as the aggregation operator. This operator provides aggregations that lie between two extreme cases of MCDM problems that lead to the use of *and* and *or* operators to combine the criteria function. OWA operator encompasses several operators since it can implement different aggregation rules by changing the order weights.

The OWA operator provides a unified framework for decision making under uncertainty, in which different decision criteria such as maximax, maximin, equally likely (Laplace), and Hurwicz's criteria are characterized by different OWA operator weights. To apply the OWA operator for decision making, a crucial issue is to determine its weights, which can be accomplished as follows.

Let $A = \{a_1, a_2, \ldots, a_n\}$ be a set of values to be aggregated, then OWA operator F is defined as

$$F(a_1, a_2, \ldots, a_n) = \mathbf{w}\mathbf{b}^T = \sum_{i=1}^{n} w_i b_i \qquad (7.3)$$

where $\mathbf{w} = \{w_1, w_2, \ldots, w_n\}$ is a weighting vector, such that $w_i \in [0,1]$ and $\sum_i w_i = 1$, and $\mathbf{b}$ is the associated ordered value vector where $b_i \in \mathbf{b}$ is the ith largest value in A.

The weights of the OWA operator are calculated using fuzzy linguistic quantifiers, which for a nondecreasing relative quantifier Q are given by

$$w_i = Q\left(\frac{i}{n}\right) - Q\left(\frac{i-1}{n}\right), \quad i = 1,\ldots,n \qquad (7.4)$$

The nondecreasing relative quantifier, Q, is defined as (Herrera et al., 2000)

$$Q(y) = \begin{cases} 0, & y < a \\ \dfrac{y-a}{b-a}, & a \le y \le b \\ 1, & y > b \end{cases} \qquad (7.5)$$

with $a,b,y \in [0,1]$ and $Q(y)$ indicating the degree to which the proportion y is compatible with the meaning of the quantifier it represents. Some nondecreasing relative quantifiers identified by terms *most*, *at least half*, and *as many as possible*, with parameters (a,b) are $(0.3,0.8)$, $(0,0.5)$, and $(0.5,1)$, respectively.

7.5 2-Tuple Fuzzy Linguistic Representation Model

The 2-tuple linguistic model, composed of a linguistic term and a real number, presented by Herrera and Martínez (2000a) is based on the concept of symbolic translation. It can be denoted as (s_g,α), where s_g represents the linguistic label of the predefined linguistic term set S_T, and α is a numerical value representing the symbolic translation. Since the 2-tuple linguistic model can express any counting of information in the universe of discourse and avoid the loss of information, it has been widely employed in decision making. This model is well suited to deal with uniformly and symmetrically distributed linguistic term sets. Moreover, the results of the Herrera and Martínez model can match the elements in the initial linguistic term set.

The process of comparison between linguistic 2-tuples is carried out according to an ordinary lexicographic order as follows (Herrera and Martínez, 2001).

Let $r_1 = (s_c, \alpha_1)$ and $r_2 = (s_d, \alpha_2)$ be two linguistic variables represented by 2-tuples:

- If $c < d$, then r_1 is smaller than r_2.
- If $c = d$, then
  - If $\alpha_1 = \alpha_2$, then r_1 and r_2 represent the same information.
  - If $\alpha_1 < \alpha_2$, then r_1 is smaller than r_2.
  - If $\alpha_1 > \alpha_2$, then r_1 is bigger than r_2.

In the following, we define a computational technique to operate with the 2-tuples without loss of information.

Definition 7.1 (Herrera and Martínez, 2000b)   Let $L = (\gamma_0, \gamma_1, \ldots, \gamma_G)$ be a fuzzy set defined in S_T. A transformation function χ that transforms L into a numerical value in the interval of granularity of $S_T, [0, G]$ is defined as

$$\chi : F(S_T) \rightarrow [0, G]$$

$$\chi(F(S_T)) = \chi(\{(s_g, \gamma_g), g = 0, 1, \ldots, G\}) = \frac{\sum_{g=0}^{G} g\gamma_g}{\sum_{g=0}^{G} \gamma_g} = \beta \qquad (7.6)$$

where $F(S_T)$ is the set of fuzzy sets defined in S_T.

Definition 7.2 (Herrera and Martínez, 2000a)   Let $S = \{s_0, s_1, \ldots, s_G\}$ be a linguistic term set and $\beta \in [0, G]$ a value supporting the result of a symbolic aggregation operation, then the 2-tuple that expresses the equivalent information to β is obtained from the following function:

$$\Delta : [0, G] \rightarrow S \times [-0.5, 0.5)$$

$$\Delta(\beta) = \begin{cases} s_g, & g = \text{round}(\beta) \\ \alpha = \beta - g, & \alpha \in [-0.5, 0.5) \end{cases} \qquad (7.7)$$

where
  round is the usual round operation
  s_g has the closest index label to β
  α is the value of the symbolic translation

Proposition 7.1 (Herrera and Martínez, 2000a)   Let $S=\{s_0,s_1,...,s_G\}$ be a linguistic term set and (s_g,α) be a 2-tuple. There is a Δ^{-1} function such that from a 2-tuple, it returns its equivalent numerical value $\beta \in [0,G] \subset \Re$. This function is defined as

$$\Delta^{-1} : S \times [-0.5,0.5) \to [0,G]$$

$$\Delta^{-1}(s_g,\alpha) = g + \alpha = \beta \tag{7.8}$$

Definition 7.3 (Herrera-Viedma et al., 2004)   Let $x=\{(s_1,\alpha_1),..., (s_G,\alpha_G)\}$ be a set of linguistic 2-tuples and $W=\{w_1,...,w_G\}$ be their associated weights. The 2-tuple weighted average $\bar{x}^w$ is computed as

$$\bar{x}^w\left[(s_1,\alpha_1),...,(s_G,\alpha_G)\right] = \Delta\left(\frac{\sum_{g=1}^{G}\Delta^{-1}(s_g,\alpha_g)\cdot w_g}{\sum_{g=1}^{G}w_g}\right)$$

$$= \Delta\left(\frac{\sum_{g=1}^{G}\beta_g\cdot w_g}{\sum_{g=1}^{G}w_g}\right) \tag{7.9}$$

Definition 7.4 (Herrera-Viedma et al., 2004; Wang, 2010)   Let $x=\{(s_1,\alpha_1),...,(s_G,\alpha_G)\}$ be a set of linguistic 2-tuples and $W=\{(w_1,\alpha_1^w),...,(w_G,\alpha_G^w)\}$ be their linguistic 2-tuple-associated weights. The 2-tuple linguistic weighted average $\bar{x}_l^w$ is calculated by the following function:

$$\bar{x}_l^w\left(\left[(s_1,\alpha_1),(w_1,\alpha_1^w)\right],...,\left[(s_G,\alpha_G),(w_G,\alpha_G^w)\right]\right)$$

$$= \Delta\left(\frac{\sum_{g=1}^{G}\beta_g\cdot\beta_{w_g}}{\sum_{g=1}^{G}\beta_{w_g}}\right) \tag{7.10}$$

with $\beta_g = \Delta^{-1}(s_g,\alpha_g)$ and $\beta_{w_g} = \Delta^{-1}(w_g,\alpha_g^w)$.

7.6 MCDM Model for Supplier Evaluation

In this section, a fuzzy multicriteria group decision-making approach integrating 2-tuple fuzzy linguistic representation model,

DEMATEL method, and QFD is proposed. In traditional QFD applications, the company has to identify its customers' expectations and their relative importance to determine the design characteristics for which resources should be allocated. When the HOQ is used in supplier selection, the company starts with the features that the outsourced product/service must possess to meet certain requirements that the company has established and then tries to identify which of the suppliers' attributes have the greatest impact on the achievement of its established objectives. The stepwise representation of the fuzzy MCDM framework is as follows:

Step 1: Construct a decision-makers committee of Z ($z = 1$, 2, ..., Z) experts. Identify the characteristics that the product being purchased must possess (CNs) in order to meet the company's needs and the criteria relevant to supplier assessment (TAs).

Step 2: Construct the decision matrices for each decision maker that denote the direct influence matrix among CNs, the fuzzy assessment to determine the CN–TA relationship scores, the degree of dependencies among TAs, and the ratings of each potential supplier with respect to each TA.

Step 3: Let the fuzzy value assigned as the CN e exerts on CN i ($i = 1$, 2, ..., m), relationship score between the ith CN and jth TA ($j = 1$, 2, ..., n), degree of dependence of the kth TA on the jth TA, and rating of the pth supplier ($p = 1$, 2, ..., P) with respect to the jth TA for the zth decision maker be $\tilde{w}_{eiz} = (w_{eiz}^1, w_{eiz}^2, w_{eiz}^3)$, $\tilde{x}_{ijz} = (x_{ijz}^1, x_{ijz}^2, x_{ijz}^3)$, $\tilde{r}_{kjz} = (r_{kjz}^1, r_{kjz}^2, r_{kjz}^3)$, and $\tilde{y}_{pjz} = (y_{pjz}^1, y_{pjz}^2, y_{pjz}^3)$, respectively. Convert $\tilde{w}_{eiz}$ into the basic linguistic scale S_T. The importance weight vector on S_T, which is denoted as $F(\tilde{w}_{eiz})$, can be represented as

$$F(\tilde{w}_{eiz}) = (\gamma(\tilde{w}_{eiz}, s_0), \gamma(\tilde{w}_{eiz}, s_1), \ldots, \gamma(\tilde{w}_{eiz}, s_8)), \quad \forall i, z \quad (7.11)$$

In this study, the label set given in Table 7.1 is used as the BLTS (Jiang et al., 2008).

Step 4: Aggregate $F(\tilde{w}_{eiz})$ using OWA operator.

Step 5: Compute β values of $F(\tilde{w}_{eiz})$ and calculate the importance weights of CNs by employing the DEMATEL method.

Table 7.1   Label Set

LABEL SET	FUZZY NUMBER
S_0	(0, 0, 0.12)
S_1	(0, 0.12, 0.25)
S_2	(0.12, 0.25, 0.37)
S_3	(0.25, 0.37, 0.50)
S_4	(0.37, 0.50, 0.62)
S_5	(0.50, 0.62, 0.75)
S_6	(0.62, 0.75, 0.87)
S_7	(0.75, 0.87, 1)
S_8	(0.87, 1, 1)

Step 6: Aggregate $\tilde{x}_{ijz}$, $\tilde{r}_{kjz}$, and $\tilde{y}_{pjz}$ using arithmetic mean operator.

Step 7: Calculate the normalized fuzzy relationships for $\alpha = 0$ and $\alpha = 1$ as

$$\left(\tilde{X}'_{ij}\right)^L_\alpha = \min \sum_{k=1}^{n} q_{ik} \left(r_{kj}\right)^L_\alpha$$

subject to

$$\sum_{k=1}^{n} q_{ik} \left(\left(r_{kj}\right)^L_\alpha + \sum_{\substack{l=1 \\ l \neq j}}^{n} \left(r_{kl}\right)^U_\alpha \right) = 1 \qquad (7.12)$$

$$\left(X_{ik}\right)^L_\alpha t \leq q_{ik} \leq \left(X_{ik}\right)^U_\alpha t, \quad k = 1, 2, \dots, n$$

$$t > 0$$

$$\left(\tilde{X}'_{ij}\right)^U_\alpha = \max \sum_{k=1}^{n} u_{ik} \left(r_{kj}\right)^U_\alpha$$

subject to

$$\sum_{k=1}^{n} u_{ik} \left(\left(r_{kj}\right)^U_\alpha + \sum_{\substack{l=1 \\ l \neq j}}^{n} \left(r_{kl}\right)^L_\alpha \right) = 1 \qquad (7.13)$$

$$\left(X_{ik}\right)^L_\alpha s \leq u_{ik} \leq \left(X_{ik}\right)^U_\alpha s, \quad k = 1, 2, \dots, n$$

$$s > 0$$

where t, s, q_{ik}, and u_{ik} are decision variables.

Step 8: Calculate the weight of each criteria $\tilde{\psi}_j = (\psi_j^1, \psi_j^2, \psi_j^3)$ for $\alpha = 0$ and $\alpha = 1$ employing

$$\left(\psi_j\right)_\alpha^U = \max \sum_{i=1}^m v_i \left(X'_{ij}\right)_\alpha^U$$

subject to

$$\lambda\left(W_i\right)_\alpha^L \leq v_i \leq \lambda\left(W_i\right)_\alpha^U, \quad i = 1, 2, \ldots, m \qquad (7.14)$$

$$\sum_{i=1}^m v_i = 1$$

$$\lambda, v_i \geq 0$$

$$\left(\psi_j\right)_\alpha^L = \min \sum_{i=1}^m v_i \left(X'_{ij}\right)_\alpha^L$$

subject to

$$\lambda\left(W_i\right)_\alpha^L \leq v_i \leq \lambda\left(W_i\right)_\alpha^U, \quad i = 1, 2, \ldots, m \qquad (7.15)$$

$$\sum_{i=1}^m v_i = 1$$

$$\lambda, v_i \geq 0$$

where λ and v_i are decision variables.

Step 9: Calculate distances from the ideal and the anti-ideal solutions (D_p^* and D_p^-, respectively) for each alternative as

$$D_p^* = \sum_{j=1}^n \frac{1}{2}\left\{\max\left(\psi_j^1 \left|y_{pj}^1 - 1\right|, \psi_j^3 \left|y_{pj}^3 - 1\right|\right) + \psi_j^2 \left|y_{pj}^2 - 1\right|\right\} \qquad (7.16)$$

$$D_p^- = \sum_{j=1}^n \frac{1}{2}\left\{\max\left(\psi_j^1 \left|y_{pj}^1 - 0\right|, \psi_j^3 \left|y_{pj}^3 - 0\right|\right) + \psi_j^2 \left|y_{pj}^2 - 0\right|\right\} \qquad (7.17)$$

Step 10: Calculate the ranking index (*RI*) of the *p*th supplier:

$$RI_p = \frac{D_p^-}{D_p^- + D_p^*} \qquad (7.18)$$

Step 11: Rank the suppliers according to RI_p values in descending order. Identify the alternative with the highest RI_p as the best supplier.

7.7 Case Study

Over the past two decades, parallel to the upsurge in the number and complexity of medical devices, the medical device industry has become intensively competitive with an increase in the number of manufacturing companies. Selecting the best medical device supplier among multiple alternatives has become one of the most critical decisions faced by purchasing managers in medical device supply chain. The performance of suppliers has a key role on cost, quality, and service in achieving customer satisfaction in the health-care industry.

In order to demonstrate the application of the proposed decision-making method to medical device supplier selection, an evaluation for epidural catheter suppliers is presented. The case study is conducted in a private hospital at the Asian side of Istanbul. The hospital operates with all major departments while including facilities such as clinical laboratories, emergency service, intensive care units, and operating room.

First, a HOQ is constructed that demonstrates the relationships between the features that epidural catheters must possess and supplier assessment criteria as well as the interactions among supplier assessment criteria. As a result of discussions with experts from the purchasing department of the hospital, nine fundamental characteristics required of epidural catheters purchased from medical suppliers (CNs) are determined. These can be listed as *cost (CN₁), kink resistant (CN₂), friction (CN₃), high tensile strength (CN₄), atraumatic tip design (CN₅), easy to thread and remove (CN₆), easy to anchor with the catheter connector (CN₇), good flow characteristics (CN₈),* and *shear resistant (CN₉).*

Nine criteria relevant to supplier assessment are identified as *product volume (TA₁), delivery (TA₂), payment method (TA₃), supply variety*

(TA₄), reliability (TA₅), experience in the sector (TA₆), earlier business relationship (TA₇), management (TA₈), and geographical location (TA₉). There are 12 suppliers who are in contact with the hospital.

The evaluation of the direct influence matrix among CNs is conducted by a committee of six decision makers $(DM_1, DM_2, DM_3, DM_4, DM_5, DM_6)$. DM_1, DM_2, and DM_3 used the linguistic term set *definitely low (DL), very low (VL), low (L), moderate (M), high (H), very high (VH),* and *definitely high (DH)* as shown in Figure 7.1, whereas the remaining three decision makers, namely DM_4, DM_5, and DM_6, preferred to use a different linguistic term set with *very low (VL), low (L), moderate (M), high (H),* and *very high (VH)* as depicted in Figure 7.2.

The β values of the direct influence matrix among CNs are given in Table 7.2.

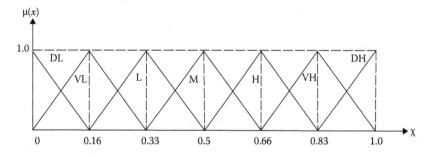

Figure 7.1   A linguistic term set where DL (0, 0, 0.16), VL (0, 0.16, 0.33), L (0.16, 0.33, 0.50), M (0.33, 0.50, 0.66), H (0.50, 0.66, 0.83), VH (0.66, 0.83, 1), and DH (0.83, 1, 1).

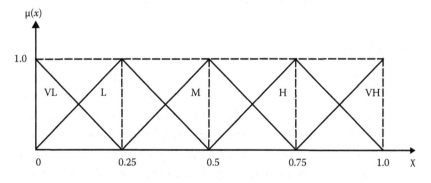

Figure 7.2   A linguistic term set where VL (0, 0, 0.25), L (0, 0.25, 0.5), M (0.25, 0.5, 0.75), H (0.5, 0.75, 1), and VH (0.75, 1, 1).

Table 7.2 β Values of the Direct Influence Matrix among CNs

	CN_1	CN_2	CN_3	CN_4	CN_5	CN_6	CN_7	CN_8	CN_9
CN_1	0.000	6.642	6.311	6.391	6.294	7.281	6.521	6.534	7.434
CN_2	7.281	0.000	6.714	6.899	6.679	7.434	1.083	4.789	7.434
CN_3	7.380	7.327	0.000	2.192	6.910	7.380	1.421	0.437	6.342
CN_4	6.968	6.279	3.482	0.000	2.731	2.296	0.054	6.642	7.242
CN_5	7.281	5.910	6.285	0.958	0.000	7.434	2.024	0.726	6.279
CN_6	7.434	6.971	6.082	1.167	6.210	0.000	1.110	0.264	6.142
CN_7	6.899	0.000	0.085	0.000	2.677	1.330	0.000	0.264	5.892
CN_8	7.434	3.998	0.057	5.696	0.759	0.000	0.000	0.000	3.760
CN_9	7.434	6.642	6.024	6.968	5.639	4.742	6.575	4.007	0.000

By employing the DEMATEL method, the weights of CNs are determined as 0.1598, 0.1337, 0.1138, 0.0993, 0.1130, 0.1130, 0.0583, 0.0709, and 0.1382, respectively. The data related to supplier selection that are provided in Table 7.3 consist of assessments of three decision makers employing linguistic variables defined in Figure 7.1.

Using Equations 7.12 through 7.15, the weights of each TA are calculated as in Table 7.4.

The distances from the ideal and the anti-ideal solutions for each alternative and the ranking index of each alternative are computed employing Equations 7.16 through 7.18 as in Table 7.5.

The rank order of the suppliers is Sup 7 > Sup 1 > Sup 4 > Sup 2 > Sup 3 > Sup 6 > Sup 9 > Sup 8 > Sup 11 > Sup 5 > Sup 10 > Sup 12. According to the results of the analysis, supplier 7 is determined as the most suitable supplier, which is followed by supplier 1, supplier 4, and supplier 2. Suppliers 10 and 12 are ranked at the bottom due to late delivery time and inadequate product volume.

7.8 Conclusion

Considering the global challenges in manufacturing environment, organizations are forced to optimize their business processes to remain competitive. In order to reach this aim, firms must work with its supply chain partners to improve the chain's total performance. Supplier's performance has a key role on cost, quality, delivery, and service in achieving the objectives of a supply chain. Hence, supplier selection is considered as one of the most critical activities of purchasing management in a supply chain.

Table 7.3 Ratings of Suppliers with respect to TAs

	TA_1	TA_2	TA_3	TA_4	TA_5	TA_6	TA_7	TA_8	TA_9
Sup 1	(VH, VH, VH)	(M, H, L)	(H, DH, H)	(VH, VH, VH)	(H, VH, VH)	(DH, DH, VH)	(H, H, M)	(H, H, H)	(M, VL, M)
Sup 2	(M, VH, M)	(H, VH, H)	(M, M, M)	(H, H, H)	(VH, H, VH)	(VH, H, DH)	(H, VH, H)	(H, H, H)	(L, L, L)
Sup 3	(M, M, M)	(H, DH, H)	(H, H, M)	(M, H, M)	(M, H, H)	(H, H, H)	(DH, VH, VH)	(M, H, M)	(H, VH, H)
Sup 4	(L, M, L)	(VH, VH, VH)	(VH, H, VH)	(L, H, L)	(H, DH, VH)	(H, VH, H)	(H, DH, H)	(H, H, VH)	(M, H, H)
Sup 5	(M, M, M)	(H, VH, H)	(H, VH, H)	(M, M, M)	(L, L, L)	(M, M, M)	(H, H, H)	(L, VL, L)	(M, M, M)
Sup 6	(H, H, H)	(H, H, H)	(VH, DH, VH)	(H, H, H)	(H, H, H)	(L, M, L)	(H, VH, H)	(H, M, H)	(M, L, M)
Sup 7	(VH, DH, VH)	(M, H, VH)	(VH, VH, DH)	(H, H, H)	(H, VH, H)	(VH, DH, H)	(H, VH, VH)	(VH, VH, VH)	(L, M, M)
Sup 8	(M, VL, L)	(M, M, M)	(DH, DH, H)	(L, L, L)	(M, H, M)	(L, H, M)	(VH, M, VH)	(M, H, M)	(VH, DH, VH)
Sup 9	(M, M, M)	(H, H, H)	(M, H, M)	(M, M, H)	(H, M, H)	(M, L, M)	(H, H, H)	(M, M, H)	(M, VH, M)
Sup 10	(L, M, L)	(L, M, VL)	(H, H, H)	(H, M, M)	(M, M, M)	(H, L, H)	(M, M, H)	(M, M, M)	(L, H, L)
Sup 11	(M, VL, M)	(L, M, VL)	(H, H, H)	(M, VL, M)	(L, L, L)	(H, M, H)	(H, VH, M)	(H, H, H)	(VH, DH, VH)
Sup 12	(DL, DL, VL)	(VL, L, DL)	(H, H, H)	(L, M, VL)	(H, H, H)	(M, M, M)	(L, L, L)	(M, VL, M)	(M, DL, L)

Table 7.4 Weights of each TA

TAs	IMPORTANCE WEIGHTS
TA_1	(0.0434, 0.0708, 0.1122)
TA_2	(0.0848, 0.1192, 0.1648)
TA_3	(0.0648, 0.0952, 0.1381)
TA_4	(0.0800, 0.1122, 0.1561)
TA_5	(0.1050, 0.1369, 0.1740)
TA_6	(0.1079, 0.1391, 0.1776)
TA_7	(0.0984, 0.1355, 0.1730)
TA_8	(0.0972, 0.1328, 0.1731)
TA_9	(0.0334, 0.0584, 0.0981)

Table 7.5 Ranking of Suppliers

SUPPLIERS	D_p^*	D_p^-	RI_p	RANK
Sup 1	0.3116	0.9437	0.7518	2
Sup 2	0.3470	0.9035	0.7225	4
Sup 3	0.3568	0.8813	0.7118	5
Sup 4	0.3275	0.9273	0.7390	3
Sup 5	0.4916	0.7248	0.5959	10
Sup 6	0.3823	0.8593	0.6921	6
Sup 7	0.2761	0.9945	0.7827	1
Sup 8	0.4438	0.7814	0.6378	8
Sup 9	0.4326	0.7898	0.6461	7
Sup 10	0.5056	0.7074	0.5832	11
Sup 11	0.4801	0.7427	0.6074	9
Sup 12	0.6291	0.5762	0.4781	12

In a medical device supply chain, identifying the most appropriate supplier among multiple alternatives is of outmost importance. In this study, a fuzzy multicriteria group decision-making algorithm is presented for medical supplier evaluation and selection. The methodology developed in this study considers QFD planning as a fuzzy multicriteria group decision tool. It enables to consider not only the impacts of relationships among the purchased product features and supplier selection criteria, but also the inner dependencies among supplier selection criteria for achieving higher satisfaction to meet company's requirements. Applying the decision framework presented here to real-world group decision-making problems in other disciplines that can be represented using HOQ matrices will be the subject of future studies.

References

Ahmady, N., M. Azadib, S.A.H. Sadeghic, and R.F. Saen. 2013. A novel fuzzy data envelopment analysis model with double frontiers for supplier selection. *International Journal of Logistics: Research and Applications*, 16 (2): 87–98.

Alinezad, A., A. Seif, and N. Esfandiari. 2013. Supplier evaluation and selection with QFD and FAHP in a pharmaceutical company. *International Journal of Advanced Manufacturing Technology*, 68: 355–364.

Alptekin, S.E. and E.E. Karsak. 2011. An integrated decision framework for evaluating and selecting e-learning products. *Applied Soft Computing*, 11: 2990–2998.

Amin, S.H. and J. Razmi. 2009. An integrated fuzzy model for supplier management: A case study of ISP selection and evaluation. *Expert Systems with Applications*, 36: 8639–8648.

Bai, C. and J. Sarkis. 2010. Integrating sustainability into supplier selection with grey system and rough set methodologies. *International Journal of Production Economics*, 124: 252–264.

Bevilacqua, M., F.E. Ciarapica, and G. Giacchetta. 2006. A fuzzy-QFD approach to supplier selection. *Journal of Purchasing & Supply Management*, 12: 14–27.

Bottani, E. and A. Rizzi. 2005. A fuzzy multi-attribute framework for supplier selection in an e-procurement environment. *International Journal of Logistics Research and Applications*, 8 (3): 249–266.

Chen, C.T., C.T. Lin, and S.F. Huang. 2006. A fuzzy approach for supplier evaluation and selection in supply chain management. *International Journal of Production Economics*, 102: 289–301.

Chuu, S.J. 2009. Group decision-making model using fuzzy multiple attributes analysis for the evaluation of advance manufacturing technology. *Fuzzy Sets and Systems*, 160 (5): 586–602.

Dursun, M. and E.E. Karsak. 2013. A QFD-based fuzzy MCDM approach for supplier selection. *Applied Mathematical Modelling*, 37: 5864–5875.

Hammami, R., C. Temponi, and Y. Frein. 2014. A scenario-based stochastic model for supplier selection in global context with multiple buyers, currency fluctuation uncertainties, and price discounts. *European Journal of Operational Research*, 233: 159–170.

Herrera, F., E. Herrera-Viedma, and L. Martínez. 2000. A fusion approach for managing multi-granularity linguistic term sets in decision making. *Fuzzy Sets and Systems*, 114 (1): 43–58.

Herrera, F. and L. Martínez. 2000a. An approach for combining linguistic and numerical information based on 2-tuple fuzzy representation model in decision-making. *International Journal of Uncertainty, Fuzziness and Knowledge-Based Systems*, 8 (5): 539–562.

Herrera, F. and L. Martínez. 2000b. A 2-tuple fuzzy linguistic representation model for computing with words. *IEEE Transactions on Fuzzy Systems*, 8 (6): 746–752.

Herrera, F. and L. Martínez. 2001. A model based on linguistic 2-tuples for dealing with multigranular hierarchical linguistic contexts in multi-expert decision-making. *IEEE Transactions on Systems, Man, and Cybernetics— Part B: Cybernetics*, 31 (2): 227–234.

Herrera-Viedma, E., F. Herrera, L. Martínez, J.C. Herrera, and A.G. López. 2004. Incorporating filtering techniques in a fuzzy linguistic multi-agent model for information gathering on the web. *Fuzzy Sets and Systems*, 148 (1): 61–83.

Jiang, Y.P., Z.P. Fan, and J. Ma. 2008. A method for group decision making with multi granularity linguistic assessment information. *Information Sciences*, 178 (4): 1098–1109.

Karsak, E.E. 2004. Fuzzy multiple objective decision making approach to prioritize design requirements in quality function deployment. *International Journal of Production Research*, 42 (18): 3957–3974.

Karsak, E.E., S. Sozer, and S.E. Alptekin. 2003. Product planning in quality function deployment using a combined analytic network process and goal programming approach. *Computers and Industrial Engineering*, 44: 171–190.

Kumar, M., P. Vrat, and R. Shankar. 2006. A fuzzy programming approach for vendor selection problem in a supply chain. *International Journal of Production Economics*, 101: 273–285.

Nazari-Shirkouhi, S., H. Shakouri, B. Javadi, and A. Keramati. 2013. Supplier selection and order allocation problem using a two-phase fuzzy multi-objective linear programming. *Applied Mathematical Modelling*, 37: 9308–9323.

Shemshadi, A., H. Shirazi, M. Toreihi, and M.J. Tarokh. 2011. A fuzzy VIKOR method for supplier selection based on entropy measure for objective weighting. *Expert Systems with Applications*, 38: 12160–12167.

Shen, L., L. Olfat, K. Govindan, R. Khodaverdi, and A. Diabat. 2013. A fuzzy multi criteria approach for evaluating green supplier's performance in green supply chain with linguistic preferences. *Resources, Conservation and Recycling*, 74: 170–179.

Shillito, M.L. 1994. *Advanced QFD: Linking Technology to Market and Company Needs*. New York: Wiley.

Tzeng, G.H., W.H. Chen, R. Yu, and M.L. Shih. 2010. Fuzzy decision maps: A generalization of the DEMATEL methods. *Soft Computing*, 14: 1141–1150.

Wang, W.P. 2010. A fuzzy linguistic computing approach to supplier evaluation. *Applied Mathematical Modelling*, 34: 3130–3141.

Wu, D.D., Y. Zhang, D. Wu, and D.L. Olson. 2010. Fuzzy multi-objective programming for supplier selection and risk modeling: A possibility approach. *European Journal of Operational Research*, 200: 774–787.

Yager, R.R. 1988. On ordered weighted averaging aggregation operators in multi-criteria decision making. *IEEE Transactions on Systems Man and Cybernetics*, 18 (1): 183–190.

8

ARC SELECTION AND ROUTING FOR RESTORATION OF NETWORK CONNECTIVITY AFTER A DISASTER

AYŞE NUR ASALY AND F. SIBEL SALMAN

Contents

8.1 Introduction and Problem Definition

Disaster management involves taking actions before and after a disaster to minimize its destructive effects. After a disaster, it is critical to reach affected areas to provide relief operations, such as search and rescue, medical services, aid delivery, and establishment of temporary shelter. Furthermore, routes should be provided for evacuation, and major gateways in the transportation system, such as airports and ports, should be accessible.

One of the outcomes of a high-impact disaster is the disruption of transportation systems, which cripples postdisaster emergency and relief activities. In the 2013 Bohol earthquake and Typhoon Haiyan, rescue workers struggled to reach ravaged towns and villages in the central Philippines (Mogato and Ng 2013). Relief operations were hampered because roads, airports, and bridges had been destroyed or were covered in wreckage. After the 2011 devastating earthquake and the resulting tsunami in northeast Japan, almost 4000 road segments, 78 bridges, and 29 railway locations were reported to be damaged (BBC News and National Police Agency of Japan 2012). Accumulated debris in the downtown of Kamaishi City, Iwate Prefecture, and a damaged arterial road (National Highway 45) virtually isolated the community from rescue efforts. About 76% of the highways in the area were closed due to damage.

This study focuses on logistics planning to ensure connectivity of road networks in the immediate disaster response stage. As experienced in many cases worldwide, roads can be severely damaged in a natural disaster. For instance, in a high-magnitude earthquake, (1) some parts of the roads may be affected as follows: blocked by building, lamppost, tree, and car debris, and deformed, distorted, and ruptured due to ground failure and liquefaction; and (2) vulnerable structures such as bridges and viaducts may collapse. Damage to other infrastructure networks, such as natural gas or drainage systems, may also cause dysfunctionality in the roads. As a result, traffic is blocked at various links of the road network, and some nodes may become unreachable.

Some of the damaged roads can be cleared or restored in a short time, whereas it may take many hours, days, or months to eliminate other types of damage. For example, after the 2011 earthquake and tsunami in Japan, Japanese road administrators immediately launched an emergency road restoration operation with the cooperation of local construction companies. The efforts concentrated on 16 routes, to establish first the vertical artery, followed by east–west routes. The operation was completed after 9 days. In general, the emergency restoration goal is to ensure connectivity of the road network and provide accessibility between people in different areas as fast as possible.

For this purpose, first, the road conditions are assessed, and time to clear/open the roads is estimated. The tasks that take too long are postponed to later stages. Then, among the remaining tasks, a subset that enables connectivity should be selected, and a fleet of machinery or vehicles routed to conduct them in the shortest time. Since some people will want to evacuate the disaster area, while others will be coming in for help, strong connectivity of the network is required.

Recently, several studies focused on upgrading a road network or improving accessibility after a disaster situation. These studies are reviewed in Section 8.2. To the best of our knowledge, the restoration of the roads after a disaster by routing a fleet of vehicles in order to ensure strong connectivity of a network has not been addressed in the literature. In this study, we define a new network optimization problem to address this topic. Since the problem combines arc routing and network design elements, it is called *Arc Routing for Connectivity Problem* (ARCP).

Before we define ARCP formally, some definitions may be useful. A connected graph contains a directed path from a node i to another node j or a directed path from j to i for every pair of nodes i and j. Otherwise, the graph is disconnected. A graph is *strongly connected* if it contains a directed path from i to j and a directed path from j to i for every pair of nodes i and j. Otherwise, the graph is disconnected in the strong sense. We define ARCP on a directed, strongly connected, and simple graph $G = (V, A)$ with nonnegative arc costs. After a natural disaster, speed of transportation is highly dependent on road and extraordinary traffic conditions, as also stated in Nolz et al. (2011). Therefore, costs are calculated in terms of estimated time instead of distance. Traversal time on an unblocked (i.e., not blocked initially) or a blocked arc after it has been unblocked (i.e., opened) is equal to c_{ij}, where (i, j) represents the arc. We refer to the fleet of emergency response machineries (including possibly lighting, drainage pump, and satellite communication vehicles) that move together as a single vehicle, which is located initially at a node d, for example, its depot or an emergency response facility. Moreover, a subset B of arcs, which are determined to be blocked according to postdisaster information on road conditions, are given such that $G_B = (V, A \backslash B)$ is disconnected in the strong sense. The set B consists of all blocked arcs, and the set R, a subset of B, represents the arcs that will be traversed and cleared by the vehicle in order to restore strong connectivity of the graph. The set R is not known in advance, and its selection is a decision in the problem. The solution identifies R and constructs a walk for the vehicle that starts at its depot. We want the walk in the solution to cover arcs in R. In other words, the arcs in the set $A \backslash B \cup R$ should induce a connected graph, G_R, on the set V.

We assume that there are $|Q|$ disconnected components in G_B, where Q is the set of disconnected components, in the strong sense. Each component in Q consists of strongly connected nodes. We partition Q into three classes: (1) components within which the nodes are strongly connected and which require at least one incoming and one outgoing arc in order to be strongly connected to the remaining graph, (2) components that require at least one outgoing but no incoming arc to be unblocked in order to be strongly connected to the remaining graph, and (3) components that require at least one incoming but no outgoing arc to be unblocked in order to be strongly connected to the remaining network. Moreover, unblocking, that is, passing through a blocked arc for the first time, results in work time in addition to its traversal time. More formally, we define the additional time of unblocking arc (i, j) as b_{ij} where $b_{ij} \geq 0$. In a walk, c_{ij} time units elapse each time an arc is traversed, and in addition, b_{ij} units elapse once for each blocked arc that is unblocked during the walk. In other words, a blocked arc is unblocked by a vehicle in its first traversal of that arc. We assume that traffic cannot flow in both directions after a blocked road is unblocked in one direction by a vehicle. Considering that allowing traffic in the reverse direction would slow down response activities, this is a reasonable assumption.

The objective is to minimize the time at which the graph becomes strongly connected. That is, by definition, there must be a path from each vertex to every other vertex in the network. In order to connect all the disconnected components, at least two arcs in opposite directions within the cutset of a component must be unblocked. Otherwise, the network cannot be strongly connected. Since we are interested in minimizing the time when the graph becomes connected, return of the vehicle to its depot is not considered. Therefore, the walk is open. We can define the objective function as min $c(W) + b(W)$, where W is walk of the vehicle; $c(W)$ is traversal time, and $c(W)$ is calculated by summing up the traversal time of arcs (in terms of c_{ij}) that are traversed by the vehicle; $b(W)$ is the total additional time (in terms of b_{ij}) of unblocking for the vehicle.

The aim of this study is to develop a solution method to the connectivity problem that generates a solution in a short time. We formulate ARCP and observe for which cases it can be solved in reasonably short time by numerical tests. Our tests are performed on instances generated considering Istanbul road network at a macro level and its vulnerability to a potential earthquake. Our analysis of the solutions over a set of scenarios provides some insights for preparedness.

The organization of this study is as follows: Section 8.2 reviews relevant studies in the literature. Section 8.3 gives computational complexity proof of the ARCP. In Section 8.4, a mixed integer programming (MIP) model for ARCP is given. Section 8.5 presents the data related to Istanbul highway network, and Section 8.6 gives the computational results. Finally, in Section 8.7, we conclude the study with a summary, some comments, and directions for future research.

8.2 Literature Review

Arc routing problems have attracted the interest of researchers for a long time and have many application areas such as delivery services and snow plowing. The problem addressed in this study falls into the class of arc routing problems. The main goal of this section is to introduce problems closely related to ARCP.

In rural postman problem (RPP), a given subset of arcs is required to be traversed at least once by a closed walk. The objective is to minimize the total travel time. RPP is NP-hard on an undirected or directed graph (Lenstra and Rinnooy Kan 1976). If the arc costs satisfy the triangle inequality, there exists a 3/2-approximation algorithm (Frederickson 1979). From this point on, the heuristic algorithm that Frederickson presents will be addressed as Frederickson's heuristic. Fernandez et al. (2003) give formulations and compare them with the former formulations from the literature. They also propose a heuristic method that is based on Frederickson's heuristic. A local search approach is applied to RPP by Groves and van Vuuren (2005). Another heuristic method is a constructive algorithm that performs local postoptimization in each step (Ghiani et al. 2006). Based on Frederickson's heuristic, Holmberg (2010) proposes heuristics using

Minimum Spanning Tree solution and postprocessing techniques. A detailed review of work before the early 1990s can be found in Eiselt et al. (1995). Akoudad and Jawab (2013) provide a recent survey that presents some variations and applications of RPP.

A variation of RPP is studied by Araoz et al. (2009). In this problem, there is no required edge to be traversed. A profit function is defined on the edges that must be taken into account for only the first time an edge is traversed. The objective is to maximize the net profit after the cost of traversing edges is deducted. They solve a relaxed model and propose a heuristic method that is based on the 3T heuristic method used in Fernandez et al. (2003).

Araoz et al. (2006) studied privatized RPP on an undirected graph and analyzed several linear systems of inequalities. In this problem, the edge profit function is similar to unblocking time in ARCP. There is a cost of traversing an edge that is paid each time the edge is traversed. Profit is collected only the first time an edge is traversed. The aim is to find a closed walk starting and ending at a depot, traversing some edges in order to maximize the total profit.

ARCP differs from the literature in several ways. In ARCP, *strong connectivity* is the main concern. Most of the other studies do not aim to ensure strong connectivity of the network. ARCP is similar to RPP, but in our problem, the set of required arcs are not known in advance, and there is no requirement for the walks to be closed. Moreover, in ARCP, after the first traversal of a blocked arc, the traversal cost changes.

In the disaster context, recently, several studies modeled upgrading the road network or improving accessibility after a disaster without considering routing. They focus on the selection of road segments that are to be upgraded or repaired. One such study is by Duque and Sörensen (2011). They investigate the case where there is a budget constraint, and there are a number of nonoperative roads that need to be repaired after a disaster situation. They assign weights to the rural towns depending on the importance of the towns. Their objective is to minimize the weighted sum of time to travel from each rural town to its closest regional center (Duque and Sörensen 2011). They find the roads to be repaired in order to have the shortest paths between node pairs. Another study is by Campbell et al. (2006), which focuses on determining the number of edges to be upgraded before a catastrophe while minimizing the maximum travel time between any source–terminal/origin–destination (s–t) pair. They use heuristic methods to solve the problem.

Only few recent studies have addressed debris removal operations in terms of selecting the order in which unblocking of the edges should be conducted. Stilp et al. (2011) model debris management as a multiperiod network expansion problem and propose efficient heuristics. Sahin et al. (2013) aim at visiting critical disaster-affected districts as quickly as possible, taking into account priority levels, traversing (if necessary) along blocked arcs by carrying out unblocking operations. They model a multiperiod mixed integer program and solve a case study. Aksu and Ozdamar (2014) consider a dynamic path-based model to identify the order of blocked links to be restored during a given time limit. The objective is to maximize the total weighted earliness of all paths' restoration completion times. ARCP differs from these problems in its objective of ensuring connectivity in shortest time.

8.3 Complexity Analysis

The problem defined in this study, namely, ARCP, is new to the arc routing literature. Therefore, we analyze the computational complexity of ARCP.

Theorem 8.1   ARCP is NP-hard.

Proof: In order to prove this theorem, we consider another NP-hard problem, RPP. We reduce RPP to ARCP.

Definition 8.1   Undirected rural postman problem (RPP).

Let $G = (V, E)$ be an undirected graph, where V is the vertex set, E is the edge set, $c_{ij}(\geq 0)$ is the cost of traversing edge $(i, j) \in E$, and $R \subseteq E$ is the set of required edges. The RPP is to determine a least cost closed walk starting from and ending at a depot, traversing each edge of R at least once. The RPP is known to be NP-hard (Lenstra and Rinnooy Kan 1976).

Now, let us consider ARCP.

Definition 8.2   Arc routing for connectivity problem (ARCP).

Let $H = (N, A)$ be a directed strongly connected graph, where N is the vertex set, A is the arc set, and $B \subseteq A$ is the set of blocked arcs. The graph induced by $A \backslash B$ is disconnected (in the strong sense). c_{ij} is the traversal time on an open arc $(i, j) \in A$, and b_{ij} is the time of unblocking edge $(i, j) \in B$ in addition to traversal time c_{ij}. ARCP finds a walk starting from its depot, traversing some of the blocked arcs in B to unblock them at the first traversal in order to connect the network. The travel time of the walk is minimized, such that the resulting graph is strongly connected. For this proof, we take an instance I of RPP and construct an instance II of ARCP by a polynomial transformation τ between them.

Definition 8.3   Transformation τ.

We define a directed and strongly connected graph H from G as follows. We replace every edge (i, j) in $E \backslash R$ with two arcs in both directions with traversal times. We take $G = (V, E)$, delete the edges in the set R and for each $(i, j) \in R$, add three new nodes i', j', and p. We define blocked arcs (i, i'), (i', i), (j, j'), and (j', j) all with traversing and additional unblocking time of 0. Moreover, between i and j, new blocked arcs (i, p), (p, i), (j, p), and (p, j) with traversal and additional unblocking time $c_{ij}/2$ and 0, respectively, are defined.

In order to transform a closed walk in I to an open walk in II, we add a dummy depot d', which is connected to the original depot d of I in the ARCP instance with two arcs in both directions, one of them blocked. Traversal and additional unblocking time on this blocked arc from d to d' is zero. The arc (d', d) that is not blocked has a high traversal time, say M. By assigning a high traversal time to this arc, we enforce the vehicle to visit d' last. The vehicle, located at d, first traverses other arcs in its walk, and then to ensure a strongly connected graph, it visits d' as the last node in its walk. It does not visit it in the early stages of its walk because then it will continue its walk to connect the remaining nodes by traversing the arc (d', d), which increases the objective value highly.

Instances I, II, and the transformation are illustrated in Figure 8.1.

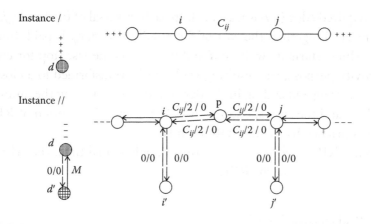

Figure 8.1 Instance I, instance II, and transformation.

Lemma 8.1 Transformation τ from I to II runs in polynomial time in terms of the size of the instance I of RPP.

Proof: For every edge in set R in I, we delete one edge and add three nodes and eight arcs. Moreover, for the depot node, one dummy depot and two arcs are added. The edges that are not required to be traversed are doubled into arcs.

Now, we need to show that we can obtain an optimal solution to I when ARCP is solved on II. The nodes i', j', p, and d' need to be visited in order to make the graph strongly connected. No matter from which direction the vehicle comes (from i to j or from j to i), it unblocks the arcs (i, i') and (j, j') to reach i' or j'. Due to the definition of ARCP, for connectivity, arcs (i', i) and (j', j) have to be unblocked as well. Moreover, node p has to be connected to the network, and unblocking one arc going out of node p and one arc coming into it is sufficient in order to ensure strong connectivity of p to the network. Possible routes for the arc segment that corresponds to a required edge can be $i-i'-i-p-j-j'-j$ or $i-i'-i-p-i-\cdots-j-j'-j$ and the reverse. In all cases, travel time of these route segments is c_{ij}. If the vehicle needs to pass through nodes i and j, it does not visit i' and j' not to increase its travel time unnecessarily. These route segments can be converted to the edge (i, j) in the RPP. Consequently, the required edge (i, j) is traversed. In each traversal of (i, p) and (j, p) (or reverse) together, the cost of the required edge, c_{ij}, is paid. Since the vehicle starts its walk in d and visits d' as the last stop for connectivity purpose, the resulting walk can be transformed to a closed walk starting and ending in the depot node d by omitting the dummy node d' and the corresponding arcs. At the end, the solution of RPP on I is reached by solving ARCP on II.

Since RPP is NP-hard and τ runs in polynomial time, the ARCP is at least as hard as the RPP.

8.4 Mathematical Model

This section presents a mathematical programming formulation of ARCP. Some properties of a feasible solution of ARCP are given as follows:

- It is necessary that arcs in a subset R of B are unblocked. The arcs in the cutsets of components are candidates to be in R. However, additional arcs may also be unblocked to reach one of these arcs in shorter time.
- In order to ensure connectivity of the graph, the total number of blocked arcs, which are unblocked in cutsets of all components has to be greater than or equal to $2(|Q| - 1)$. Otherwise, connectivity cannot be ensured. In other words, in each component's cutset, at least two arcs that are in opposite directions must be open. This property is necessary for a solution to be feasible, but it is not sufficient for optimality.

In order to ensure connectivity and continuity of the walk, we define flow variables f_{ij} for each arc. For the depot, there is an amount of supply depending on the number of nodes that are visited by the vehicle. Similarly, for each component, there is unit demand so that each component can receive flow and the graph becomes connected at the end. Then, to prevent flows on an arc that is not traversed, we relate flow variables with x_{ij}, which shows the number of times an arc (i, j) is traversed. Flow variables are defined as real numbers, however, due to unimodularity property; they take integer values because flow variables in the constraints have integer coefficients. Moreover, we add a dummy sink node and force the vehicle to end its tour at this sink node $(n + 1)$. For connectivity, we include cutset constraints. The details can be seen in the upcoming paragraphs.

8.4.1 Sets, Indices, and Input Parameters

i, j: Indices of the vertices

$n + 1$: Index of the dummy sink node

V: Set of vertices: $1, ..., n$

A: Set of arcs

B: Set of blocked arcs

d: Index for the depot

D: Set of possible depots

q: Index of the components

Q: Set of disconnected components

S: Set of all subsets of components within which the nodes are strongly connected

s: Index of elements of S

Y^+: Set of all subsets of components that require at least one outgoing arc but no incoming arc to be unblocked in order to be strongly connected to the remaining graph

Y^-: Set of all subsets of components that require at least one incoming arc but no outgoing arc to be unblocked in order to be strongly connected to the remaining graph

y: Index of the components

M: A nonnegative scalar with large enough value

8.4.2 Decision Variables

x_{ij}: Number of times that the vehicle traverses arc (i, j)

z_{ij}: Binary variable indicating if blocked arc (i, j) is unblocked

f_{ij}: Flow variable on arc (i, j)

v_i: Number of times the vehicle visits node i

The MIP model for ARCP determines an open walk such that the disconnected components in the network are connected after unblocking a subset of the blocked arcs. The walk traverses a subset of the arcs in B, say R, so that the graph $G' = (V, A\backslash B \cup R)$ is connected. The model that solves ARCP gives a strongly connected graph. We explain the objective function and constraints group by group as follows.

8.4.3 Objective Function

Constraint (8.1) represents the objective function that minimizes the total time spent by the vehicle until the network becomes strongly connected:

$$\text{Minimize} \sum_{(i,j)\in A} c_{ij}x_{ij} + \sum_{(i,j)\in B} b_{ij}z_{ij} \tag{8.1}$$

8.4.4 Vehicle Balance Equations

Constraints (8.2) through (8.5) are vehicle balance equations. Constraint (8.2) ensures that the vehicle starts the tour at the depot vertex where it is positioned. Constraint (8.3) balances arrivals and departures for a nondepot node i. Constraint (8.4) forces the walk to end in the sink node. There is only one visit to the sink node and no return. The latter case is satisfied by constraint (8.5). The vehicle leaves the depot and its component, and does not return there if it will not visit another disconnected component by passing through its own component:

$$\sum_{j\in V \cup \{(n+1)\}} \left(x_{dj} - x_{jd}\right) = 1, \quad d \in D \tag{8.2}$$

$$\sum_{j\in V \cup \{(n+1)\}} \left(x_{ij} - x_{ji}\right) = 0, \quad \forall i \in V \setminus D \tag{8.3}$$

$$\sum_{j\in V} x_{j(n+1)} = 1 \tag{8.4}$$

$$x_{(n+1)i} = 0, \quad \forall i \in V \tag{8.5}$$

8.4.5 Constraints That Relate Variables x_{ij} and z_{ij}

Constraint (8.6) shows for a blocked arc that if it is unblocked, then it is also traversed. We assume a blocked arc becomes open in both directions whenever the vehicle unblocks it in one direction. This assumption can be meaningful because in disaster situations, roads have to be used in both directions in order to reach disaster areas and deliver aid. Constraint (8.7) prevents the vehicle traversing a blocked arc if it is not unblocked. If an arc (i, j) is unblocked, it can be traversed by the vehicle at most $2(|Q| - 1)$ times. The vehicle connects one component each time it traverses the same arc by unblocking one arc going out of the subset of component and one arc coming into it. Therefore, we multiply this value by 2. Except the component that it is deployed, there are $(|Q| - 1)$ components in total to be connected; thus, the scalar in this constraint takes the value of $2(|Q| - 1)$:

$$x_{ij} \geq z_{ij}, \quad \forall (i, j) \in B \tag{8.6}$$

$$x_{ij} \leq 2(|Q|-1)z_{ij}, \quad \forall (i, j) \in B \tag{8.7}$$

8.4.6 Flow Balance Equations

For connectivity of the nodes in the vehicle's walk, we define flow variables f_{ij} or each arc that it passes through. For the depot vertex, the net flow into it is the total number of visits to all vertices except the depot (as seen in constraint [8.8]). For the other vertices, it is equal to the number of visits to the corresponding node (as seen in constraint [8.9]). In other words, the vehicle leaves one unit of flow each time it visits a node. Constraint (8.10) prevents backward flow from the sink node to any other node. Constraint (8.11) requires that the walk ends in sink node by sending one unit of flow to the sink node:

$$\sum_{j:(i,j)\in A,\{i,j\}\in V\cup\{(n+1)\}} \left(f_{ij} - f_{ji} \right) = -v_i, \quad \forall i \in V \cup \{(n+1)\} \setminus D \tag{8.8}$$

$$\sum_{j\in V\cup\{(n+1)\}} \left(f_{dj} - f_{jd} \right) = \sum_{i\in V\cup\{(n+1)\}\{d\}} v_i, \quad d \in D \tag{8.9}$$

$$f_{(n+1)j} = 0, \quad \forall j \in V \tag{8.10}$$

$$\sum_{j\in V} f_{j(n+1)} = 1 \tag{8.11}$$

8.4.7 *Constraints That Relate Variables f_{ij} and x_{ij}*

Constraint (8.12) does not allow flow on an arc unless it is traversed. Constraint (8.13) shows that if an arc is traversed, then there must be a positive amount of flow passing through it:

$$f_{ij} \le M x_{ij}, \quad \forall (i,j) \in A, \{i,j\} \in V \cup \{(n+1)\} \qquad (8.12)$$

$$f_{ij} \ge x_{ij}, \quad \forall (i,j) \in A, \{i,j\} \in V \cup \{(n+1)\} \qquad (8.13)$$

8.4.8 *Component Connectivity Constraints*

For component connectivity, (8.14) and (8.15) require at least one arc into and one arc out of each subset of components within which the nodes are strongly connected to be unblocked. Similarly, with constraints (8.16) and (8.17), for connectivity of the components in sets Y^+ and Y^-, at least one arc into and one arc out of each subset are unblocked. As a result, the graph becomes strongly connected:

$$\sum_{(i,j) \in \delta^+(s)} z_{ij} \ge 1, \quad \forall s \subset S \qquad (8.14)$$

$$\sum_{(i,j) \in \delta^-(s)} z_{ij} \ge 1, \quad \forall s \subset S \qquad (8.15)$$

$$\sum_{(i,j) \in \delta^+(y)} z_{ij} \ge 1, \quad \forall y \subset Y^+ \qquad (8.16)$$

$$\sum_{(i,j) \in \delta^-(y)} z_{ij} \ge 1, \quad \forall y \subset Y^- \qquad (8.17)$$

8.4.9 *Constraints That Define the Variables*

Constraints (8.18) through (8.20) are integrality constraints, whereas constraints (8.21) are binary constraints. Constraints (8.22) state that flow variables are nonnegative real numbers:

$$x_{ij}, x_{ji} \in \mathbb{Z}_+, \quad \forall (i,j) \in A, \{i,j\} \in V \qquad (8.18)$$

$$x_{i(n+1)} \in \mathbb{Z}_+, \quad \forall i \in V \qquad (8.19)$$

$$v_i \in \mathbb{Z}_+, \quad \forall i \in V \qquad (8.20)$$

$$z_{ij} \in \mathbb{B}, \quad \forall(i,j) \in B \qquad (8.21)$$

$$f_{ij}, f_{ji} \in \mathbb{R}_+, \quad \forall(i,j) \in A \qquad (8.22)$$

8.5 Data Acquisition and Generation

For computational experiments, we constructed a network of Istanbul that is obtained by considering province centers and real road distances. By using Google Maps, we identified strategically important locations such as province centers and provinces that have hospitals, disaster coordination centers, ports, airports, bus terminals, and bridges. Possible depot points are given in Table 8.1. Depot points are determined according to the locations related to highway maintenance, the locations that may have machinery, for example, cranes and trucks. There are 74 nodes including 38 province centers and 34 populated districts (see Figure 8.2). In total, there are 360 links (720 arcs) (see Figure 8.3). Arcs are created between neighbors, and arc traversal times are determined by using road distances given in Table 8.2, which are calculated using Google Maps. We converted road distances into time (in hours) assuming an average 50 km/h speed for the vehicle.

Table 8.1 Possible Locations of Depots

	PROVINCE	NODE
Disaster Coordination Center	Kağıthane	23
GDH Division of Machinery Supply	Maltepe	29
GDH Division of Road Maintenance and Repair	Kartal	32
GDH Division of Road Maintenance and Repair	Edirnekapı/Eyüp	15
GDH Regional Division of Maintenance and Operations	Kavacık	27
GDH Regional Division of Maintenance and Operations	Kurtköy/Pendik	36
General Directorate of Highways (GDH)	Kağıthane	23
Istanbul Metropolitan Municipality	Fatih	19
Istanbul Metropolitan Municipality—additional building	Merter/Güngören	17

Figure 8.2 Nodes on Istanbul map from Google Earth.

Ten scenarios with different sets of blocked roads are generated by referring to the latest earthquake risk map of Istanbul reported by the Japan International Cooperation Agency and Istanbul Metropolitan Municipality in a 2002 study (The Japan International Cooperation Agency [JICA]; Istanbul Metropolitan Municipality 2002). We classified the roads into three based on the earthquake risk map: high-risk roads (see Table 8.3 for high-risk roads), low-risk roads (see Table 8.4 for low-risk roads), and the remaining ones. More roads are picked to be blocked in high-risk area than in low-risk area, but within each risk level, blocked roads are selected randomly. In this way, three to six disconnected components are formed. The number of disconnected components and the number of blocked roads in each scenario are given in Table 8.5.

For each scenario, two instances with high and low unblocking times are generated. Unblocking time of an arc is set proportional to its traversal time, that is, $b_{ij} = \alpha\, c_{ij}$. The factor α is generated randomly as follows. First, blocked roads are classified into high-, medium-, and low-damage groups randomly with probabilities listed in Table 8.6. In high–unblocking time case, high-damage roads are more likely and low-damage roads are less likely. For example, around 60% of the blocked arcs would have high damage, while 10% would have low damage. The factor α has a uniform distribution and takes values between (10, 50), (5, 10), and (2, 5) for high-, medium-, and low-damage groups, respectively.

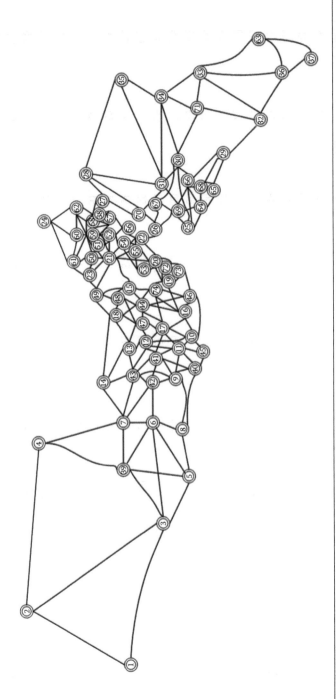

Figure 8.3   The network representing the main roads in Istanbul.

Table 8.2 Real Road Distances

ORIGIN NODE	DESTINATION NODE	DISTANCE	ORIGIN NODE	DESTINATION NODE	DISTANCE	ORIGIN NODE	DESTINATION NODE	DISTANCE
1	2	39.0	22	56	3.0	48	44	2.5
1	3	39.0	23	21	3.0	48	49	4.5
2	1	39.0	23	49	5.0	49	15	4.5
2	3	18.0	23	50	1.2	49	18	4.5
2	4	45.0	23	61	5.5	49	21	5.5
3	1	39.0	24	60	8.0	49	23	5.0
3	2	18.0	24	61	11.0	49	48	4.5
3	5	12.0	24	62	10.0	50	21	2.0
3	6	12.0	25	30	7.5	50	23	1.2
3	39	11.0	25	64	6.5	50	51	2.5
4	2	45.0	25	66	10.0	50	59	3.5
4	7	25.0	25	68	10.0	50	61	6.0
4	39	30.0	25	69	3.5	51	21	2.5
5	3	12.0	26	22	9.5	51	50	2.5
5	6	9.5	26	67	3.0	51	52	1.5
5	8	11.0	26	69	5.0	51	59	2.5
5	39	9.5	26	70	2.5	52	21	2.5
6	3	12.0	27	28	6.5	52	51	1.5
6	5	9.5	27	38	6.5	52	54	3.0
6	7	5.5	27	70	10.0	52	56	2.5
6	8	10.0	28	27	6.5	52	59	1.9

6	39	4.0	28	31	17.0	53	20	2.0
6	42	18.0	28	35	21.0	53	54	4.0
7	4	25.0	28	70	17.0	53	55	2.5
7	6	5.5	29	32	6.5	54	21	2.5
7	14	26.0	29	65	5.0	54	22	2.0
7	39	7.0	29	66	8.0	54	52	3.0
7	42	10.0	29	68	6.5	54	53	4.0
7	43	19.0	30	25	7.5	54	55	3.0
8	5	11.0	30	31	5.0	55	20	1.6
8	6	10.0	30	34	28.0	55	21	3.5
8	9	9.0	30	66	4.0	55	22	4.0
8	40	12.0	30	67	5.5	55	53	2.5
9	8	9.0	30	71	15.0	55	54	3.0
9	11	6.0	31	28	17.0	56	22	3.0
9	40	8.0	31	30	5.0	56	52	2.5
9	41	3.0	31	34	16.0	56	57	3.5
9	42	14.0	31	35	9.5	57	38	0.5
9	43	11.0	31	67	4.5	57	56	3.5
10	11	5.0	31	69	5.0	57	58	2.5
10	12	7.0	32	29	6.5	57	59	3.0
10	16	3.5	32	33	16.0	58	38	1.7
10	17	5.0	32	36	4.5	58	57	2.5
10	45	3.5	32	71	13.0	58	59	1.5

(Continued)

Table 8.2 (*Continued*)   Real Road Distances

ORIGIN NODE	DESTINATION NODE	DISTANCE
10	46	4.5
11	9	6.0
11	10	5.0
11	12	6.5
11	17	5.0
11	40	8.0
11	41	5.0
11	45	3.5
11	10	7.0
12	11	6.5
12	13	3.0
12	17	4.5
12	41	5.0
12	43	10.0
12	47	12.0
13	12	3.0
13	14	11.0
13	18	8.5
13	43	11.0
13	47	8.0
14	7	22.0
33	32	16.0
33	34	5.0
33	36	13.0
33	63	7.5
33	71	5.0
34	30	28.0
34	31	16.0
34	33	5.0
34	35	7.0
34	71	5.0
35	28	21.0
35	31	9.5
35	34	7.0
36	32	4.5
36	33	13.0
36	37	11.0
36	63	8.0
37	36	11.0
37	63	13.0
38	27	6.5
38	57	0.5
58	62	6.0
59	21	3.5
59	50	3.5
59	51	2.5
59	52	1.9
59	57	3.0
59	58	1.5
59	60	6.0
59	61	7.0
59	62	8.0
60	24	8.0
60	59	6.0
60	61	3.0
60	62	4.0
61	23	5.5
61	24	11.0
61	50	6.0
61	59	7.0
61	60	3.0
62	24	10.0
62	38	4.0

14	13	11.0	38	58	1.7	62	58	6.0
14	18	10.0	38	62	4.0	62	59	8.0
14	43	13.0	39	3	11.0	62	60	4.0
15	18	2.5	39	4	30.0	63	33	7.5
15	21	7.5	39	5	9.5	63	36	8.0
15	44	3.0	39	6	4.0	63	37	13.0
15	48	1.6	39	7	7.0	64	25	6.5
15	49	4.5	40	8	12.0	64	65	5.5
15	74	2.0	40	9	8.0	64	66	6.0
16	10	3.5	40	11	8.0	64	68	4.5
16	17	4.5	40	41	8.5	64	69	6.0
16	19	6.5	40	45	4.0	65	29	5.0
16	44	9.0	41	9	3.0	65	64	5.5
16	46	1.9	41	11	5.0	65	68	2.0
16	74	7.5	41	12	5.0	66	25	10.0
17	10	5.0	41	17	11.0	66	29	8.0
17	11	5.0	41	40	8.5	66	30	4.0
17	12	4.5	41	42	8.0	66	64	6.0
17	16	4.5	41	43	5.5	66	68	3.5
17	41	11.0	42	6	18.0	66	69	7.5
17	44	4.5	42	7	10.0	67	26	3.0
17	47	7.0	42	9	14.0	67	30	5.5
17	74	7.0	42	41	8.0	67	31	4.5

(Continued)

Table 8.2 (Continued)   Real Road Distances

ORIGIN NODE	DESTINATION NODE	DISTANCE	ORIGIN NODE	DESTINATION NODE	DISTANCE	ORIGIN NODE	DESTINATION NODE	DISTANCE
18	13	8.5	42	43	2.5	67	69	3.5
18	14	10.0	43	7	19.0	67	70	2.0
18	15	2.5	43	9	11.0	68	29	6.5
18	44	3.0	43	12	10.0	68	64	4.5
18	47	8.0	43	13	11.0	68	65	2.0
18	48	1.9	43	14	14.0	68	66	3.5
18	49	4.5	43	41	5.5	69	25	3.5
19	16	6.5	43	42	2.5	69	26	5.0
19	20	3.0	44	15	3.0	69	31	5.0
19	72	3.0	44	16	9.0	69	64	6.0
19	73	3.0	44	17	4.5	69	66	7.5
19	74	3.0	44	18	4.0	69	67	3.5
20	19	3.0	44	46	7.0	70	26	2.5
20	21	5.0	44	47	5.0	70	27	18.0
20	22	4.0	44	48	2.5	70	28	17.0

20	53	1.8	44	74	3.5	70	67	2.0
20	55	1.6	45	10	3.5	71	30	15.0
20	72	1.8	45	11	3.5	71	32	13.0
21	15	10.0	45	40	4.0	71	33	5.0
21	20	5.0	46	10	4.5	71	34	5.0
21	23	3.0	46	16	1.9	72	19	3.0
21	49	5.5	46	44	7.0	72	20	1.8
21	50	2.0	46	73	7.0	72	73	1.3
21	51	2.5	46	74	7.0	73	19	3.0
21	52	2.5	47	12	12.0	73	46	7.0
21	54	2.5	47	13	9.5	73	72	1.3
21	55	3.5	47	17	7.0	74	15	2.0
21	59	3.5	47	18	8.0	74	16	7.5
22	20	4.0	47	44	5.0	74	17	7.0
22	26	9.5	47	48	6.5	74	19	3.0
22	54	2.0	48	15	1.6	74	44	3.5
22	55	4.0	48	18	2.5	74	46	7.0

Table 8.3 High-Risk Roads

EUROPE		ASIA	
I	J	I	J
3	5	25	30
5	8	25	69
8	9	25	64
8	40	25	66
9	11	29	32
9	40	30	71
10	11	32	33
10	17	32	36
10	16	32	71
10	46	33	36
16	46	33	63
16	19	36	63
16	74	36	37
19	73	37	63
19	72	64	68
19	20	64	65
20	22	64	66
40	45	64	69
40	11	65	68
45	10	65	29
72	20	66	29
72	73	66	30
		66	69
		68	66
		68	29
		71	33

8.6 Computational Experiments and Results

Effects of the following parameters on computational performance and objective value are analyzed: (1) degree of damage (i.e., high and low b_{ij} cases) and (2) location of the depot. To solve the models, CPLEX 12.5 was run as a multithreaded application (using GAMS 24.0 and a computer with two 3.30 GHz processors, 32 GB RAM under 64-bit operating system).

The results of high and low blocking time cases (with 23 as the depot node) can be seen in Table 8.7. All scenarios are solved to optimality in a short time (at most 114 s and in less than a minute for all

Table 8.4 Low-Risk Roads

						EUROPE						ASIA	
I	J	I	J	I	J	I	J	I	J	I	J	I	J
1	3	13	47	21	50	43	41	52	51	61	24	26	67
1	2	13	18	23	61	43	9	52	56	61	60	26	69
2	4	14	13	23	50	44	16	53	55	61	59	27	28
2	3	14	18	23	21	44	74	53	54	62	24	27	70
3	39	15	74	38	27	44	15	53	20	62	58	28	31
4	7	15	21	38	57	44	48	54	52	62	38	28	35
4	39	15	49	39	7	47	44	54	22	74	19	31	35
6	7	17	16	39	6	47	18	55	22	74	46	31	30
6	42	17	74	39	5	47	48	55	54			31	69
6	5	17	44	41	40	47	12	55	20			31	34
6	8	17	47	41	11	48	15	56	22			34	33
6	3	17	16	41	12	48	49	57	56			34	71
7	42	18	49	41	17	48	18	58	57			34	30
7	43	18	44	41	9	49	23	58	38			35	34
7	14	20	55	42	41	49	21	59	57			67	31
11	17	21	20	42	43	50	51	59	60			67	30
12	11	21	55	42	9	50	59	59	62			67	69
12	10	21	54	43	14	50	61	59	58			70	28
12	17	21	52	43	12	51	59	60	24			70	67
13	12	21	51	43	13	52	59	60	62			70	26

Table 8.5 Scenarios

SCENARIOS	NUMBER OF BLOCKED ARCS	NUMBER OF DISCONNECTED COMPONENTS
1	30	3
2	32	3
3	40	4
4	42	3
5	52	4
6	60	4
7	76	5
8	80	6
9	82	6
10	84	5

Table 8.6   Damage Level, Probabilities, α, Classification of Blocked Roads

	HIGH–UNBLOCKING TIME CASE			LOW–UNBLOCKING TIME CASE		
DAMAGE	HIGH	MEDIUM	LOW	HIGH	MEDIUM	LOW
Probability	0.6	0.3	0.1	0.1	0.3	0.6
Distribution of α	$U(10, 50)$	$U(5, 10)$	$U(2, 5)$	$U(10, 50)$	$U(5, 10)$	$U(2, 5)$

Table 8.7   Effect of Degree of Damage and Computational Results

SCENARIOS	NUMBER OF BLOCKED ARCS	NUMBER OF DISCONNECTED COMPONENTS	HIGH COST			LOW COST		
			OBJECTIVE (H)	LB (H)	TIME (S)	OBJECTIVE (H)	LB (H)	TIME (S)
1	30	3	4.5	4.5	39.0	2.5	2.5	30.0
2	32	3	3.0	3.0	8.0	2.4	2.4	10.0
3	40	4	8.5	8.5	8.0	5.1	5.1	9.0
4	42	3	6.4	6.4	6.0	3.4	3.4	9.0
5	52	4	3.3	3.3	8.0	2.7	2.7	8.0
6	60	4	6.1	6.1	13.0	3.4	3.4	9.0
7	76	5	7.3	7.3	50.0	5.5	5.5	114.0
8	80	6	11.2	11.2	27.0	5.9	5.9	27.0
9	82	6	9.1	9.1	23.0	5.9	5.9	17.0
10	84	5	8.2	8.2	14.0	3.5	3.5	21.0
1. *Average*			6.7	6.7	19.6	4.0	4.0	25.4

but one). This shows that for problems with the tested size, the goal of solving the problem very quickly is achieved. Low–unblocking time case gives 23% higher runtime on average, compared to the high–unblocking time case. As the number of components and blocked arcs increases, the effect of damage level on runtime can be observed better. We can conclude that when unblocking times decrease, solution time increases since the decision of which arcs to unblock gets more difficult, and both connectivity and routing decisions affect the solution value strongly.

In order to evaluate the effect of the location of the depot on the runtime and objective value, we picked several different nodes as the depot and solved the model with high unblocking times. Nodes 15 and 23 located in European side of Istanbul and nodes 27, 29, and 32 in Asian side are picked one by one as the depot. It is possible to consider other nodes, but we picked these for demonstration purposes. Table 8.8 shows the results for all scenarios. When the depot is at node 15, 27, 29, or 32, all scenarios are solved even faster, and the

Table 8.8   Effect of Location of the Depot on the Solution

SCENARIOS	OBJECTIVE (H)	LB (H)	TIME (S)	OBJECTIVE (H)	LB (H)	TIME (S)
	Depot ID: 15			*Depot ID: 23*		
1	4.3	4.3	10.0	4.5	4.5	39.0
2	2.8	2.8	6.0	3.0	3.0	8.0
3	8.6	8.6	8.0	8.5	8.5	8.0
4	6.2	6.2	6.0	6.4	6.4	6.0
5	3.2	3.2	6.0	3.3	3.3	8.0
6	5.9	5.9	9.0	6.1	6.1	13.0
7	7.2	7.2	22.0	7.3	7.3	50.0
8	11.1	11.1	29.0	11.2	11.2	27.0
9	8.9	8.9	13.0	9.1	9.1	23.0
10	8.2	8.2	11.0	8.2	8.2	14.0
Average	6.6	6.6	12.0	6.7	6.7	19.6
	Depot ID: 27			*Depot ID: 29*		
1	4.5	4.5	15.0	4.0	4.0	6.0
2	3.2	3.2	9.0	2.7	2.7	5.0
3	8.6	8.6	8.0	8.4	8.4	8.0
4	7.7	7.7	12.0	7.0	7.0	6.0
5	3.5	3.5	8.0	3.0	3.0	6.0
6	6.3	6.3	15.0	6.1	6.1	8.0
7	7.4	7.4	32.0	6.5	6.5	8.0
8	11.2	11.2	39.0	10.4	10.4	11.0
9	9.3	9.3	13.0	8.8	8.8	11.0
10	8.2	8.2	14.0	7.8	7.8	9.0
Average	7.0	7.0	16.5	6.5	6.5	7.8
	Depot ID: 32					
1	4.1	4.1	7.0			
2	2.9	2.9	6.0			
3	8.5	8.5	8.0			
4	7.1	7.1	6.0			
5	3.1	3.1	6.0			
6	5.9	5.9	19.0			
7	6.3	6.3	9.0			
8	10.3	10.3	12.0			
9	9.0	9.0	18.0			
10	8.4	8.4	11.0			
Average	6.6	6.6	10.2			

objective value does not change much. Choosing node 29 as the depot seems rational since it gives a better solution in terms of objective value and runtime. The reason for this performance may be explained as follows. The network has a rectangular shape, and node 29 resides in bottom-right corner. Node 32 has a similar position. The other depots are in a more central position with respect to the layout of the network. Starting from a central location may result in traversals back and forth to the component in the center. Therefore, the travel time may be longer.

8.7 Conclusions

In this study, we introduced the ARCP, which is applicable for restoring network connectivity after a disaster. The aim is to make the disconnected graph strongly connected in the shortest time by unblocking some of the blocked roads. The responsible team leaves the depot and unblocks selected roads with an unblocking time that is spent only for the first time the blocked road is traversed. We show that ARCP is NP-hard and develop an MIP formulation that can be solved quickly for instances with realistic size. This can be contributed to the fact that arc routing part of the problem is handled efficiently by sending flows. To the best of our knowledge, this is new in the arc routing literature.

To generate test data, Istanbul highway network is used. Ten different scenarios with differing blocked arcs are constructed by selecting links in high-risk areas in existing earthquake scenarios. Two levels of damage are defined, and unblocking times are calculated accordingly, leading to 20 test instances. While MIP is solved in at most 2 min in all of 20 instances, we observe that high–unblocking time cases are easier to solve than low–unblocking time cases. Changing the location of the depot does not affect the objective value and runtime much in our instances. We expect the solution of larger instances to take longer. However, having a single fleet traverse a wide area with many arcs will not provide an efficient solution. Instead, covering an area by multiple teams would be the way to go in a disaster situation.

In order to analyze the computational performance of the model, a more extensive numerical study would be required. Here, we have used only a single network with 20 postdisaster scenarios. Instead of the Istanbul map, a smaller network can be used. For example, different provinces in Istanbul can be taken, and a more detailed road network with shorter distances can be generated for each one. Then, having a single vehicle or fleet responsible from each region would reduce the completion time for connectivity of the larger network.

When the network gets larger, having multiple vehicles becomes necessary. The timing of the vehicles becomes an issue since a vehicle may need to wait while another works on an arc. A heuristic approach to solve the multivehicle case quickly can be by partitioning the graph and solving the ARCP exactly for each partition. Clearly, the quality of the solutions would depend on the partitioning step.

This line of research can be extended in several directions in future research. The problem can be defined on an undirected graph. If connecting the entire network takes too long time, connectivity to certain nodes such as supply points, hospitals, and airports can be prioritized. The objective would change to connecting given origin–destination pairs. This modification can be handled by multicommodity flows for connectivity.

References

Akoudad, K. and F. Jawab. Recent survey on bases routing problems: CPP, RPP and CARP. *International Journal of Engineering Research & Technology (IJERT)* 2 (2013): 3652–3668.

Aksu, D. T. and L. Ozdamar. A mathematical model for post-disaster road restoration: Enabling accessibility and evacuation. *Transportation Research* 61, Part E (2014): 56–67.

Araoz, J., E. Fernandez, and O. Meza. Solving the prize-collecting rural postman problem. *European Journal of Operational Research* 196 (2009): 886–896.

Araoz, J., E. Fernandez, and C. Zoltan. Privatized rural postman problems. *Computers and Operations Research* 33 (2006): 3432–3449.

BBC News and National Police Agency of Japan. Japan quake: Loss and recovery in numbers. *BBC News* March 11, 2012. http://www.bbc.com/news/world-asia-17219008 (accessed August 13, 2014).

Campbell, A., T. Lowe, and L. Zhang. Upgrading arcs to minimize the maximum travel. *Networks* 47 (2006): 72–80.

Duque, P. M. and K. Sörensen. GRASP metaheuristic to improve accessibility after. *OR Spectrum* 33 (2011): 525–542.

Eiselt, H. A., M. Gendreau, and G. Laporte. Arc routing problems, part II: The rural postman problem. *Operations Research* 43 (1995): 399–414.

Fernandez, E., O. Meza, R. Garfinkel, and M. Ortega. On the undirected rural postman problem: Tight bounds based on a new formulation. *Informs* 51 (2003): 281–291.

Frederickson, G. N. Approximation algorithms for some postman problems. *Journal of ACM* 26 (1979): 538–554.

Ghiani, G., D. Lagana, and R. Musmanno. A constructive heuristic for the undirected rural postman problem. *Computers and Operations Research* 33 (2006): 3450–3457.

Groves, G. V. and J. H. van Vuuren. Efficient heuristics for the rural postman problem. *ORiON* 21 (2005): 33–51.

Holmberg, K. Heuristics for the rural postman problem. *Computers and Operations Research* 37 (2010): 981–990.

Lenstra, J. K. and A. H. G. Rinnooy Kan. On general routing problems. *Networks* 6 (1976): 273–280.

Mogato, M. and R. Ng. Philippine typhoon survivors beg for help as rescuers struggle, *in.reuters.com.* November 11, 2013. http://in.reuters.com/article/2013/11/11/philippines-typhoon-haiyan-idINDEE9A802120131111 (accessed August 13, 2014).

Nolz, P. C., F. Semet, and K. F. Doerner. Risk approaches for delivering disaster relief supplies. *OR Spectrum* 33 (2011): 543–569.

Sahin, H., O. E. Karasan, and B. Y. Kara. On debris removal during the response phase. *Proceedings of the INOC 2013*, Tenerife, Spain, May 20–22, 2013.

Stilp, K., J. A. Carbajal, O. Ergun, P. Keskinocak, and M. Villareal. Managing debris operations. *2011 Health and Logistics Conference Poster Session*, Georgia Tech. Supply Chain & Logistics Institute Center for Humanitarian Logistics, Atlanta, GA, March 2–3, 2011.

The Japan International Cooperation Agency (JICA); Istanbul Metropolitan Municipality. *The Study on a Disaster Prevention/Mitigation Basic Plan in Istanbul Including Microzonation in the Republic of Turkey.* Istanbul Metropolitan Municipality, Istanbul, Turkey, 2002.

9

FEASIBILITY STUDY OF SHUTTLE SERVICES TO REDUCE BUS CONGESTION IN DOWNTOWN IZMIR

ERDİNÇ ÖNER, MAHMUT ALİ GÖKÇE,
HANDE ÇAKIN, AYLİN ÇALIŞKAN,
EZGİ KINACI, GÜRKAN MERCAN,
EZEL İLKYAZ, AND BERİL SÖZER

Contents

9.1 Introduction

Experiencing traffic congestion becomes inevitable for most people living in large cities, especially during rush hours. The growth of population and employment, especially in city centers, is the main reason behind this traffic congestion.

According to the 2012 Urban Mobility Report for United States prepared by Texas Transportation Institute of Texas A&M University, the estimated annual travel delay is increasing drastically. The annual travel delay was 1.1 billion hours in 1982 and reached

5.5 billion hours in 2011. Moreover, it was recorded that while the amount of CO_2 produced during congestion was 10 million lb in 1982, it increased to 47 million lb in 2000 and then to 56 million lb in 2011 (Schrank et al., 2012). An efficient public transport system can smooth traffic, reduce people's travel time, and help reduce environmental pollution. Based on the 2012 Urban Mobility Report, extending public transportation has significant savings. While its contribution to savings in yearly travel delay was 409 million hours in 1982, it increased to 865 million hours in 2011. In addition, the use of public transportation resulted in an annual congestion cost saving of $8.0 billion in 1982 and $20.8 billion in 2011.

9.1.1 Statement of the Problem

Izmir is the third most crowded city with a population of 4.1 million, and it also has the second largest port in Turkey. Public bus transportation activities in Izmir are managed by ESHOT (Izmir Public Transportation Authority) in five main districts (Figure 9.1). ESHOT launched *Smart Ticketing* system for public bus transportation on March 15, 1999. Following that progress, smart ticketing system was also integrated to the subway, suburban railway, and sea transportations. ESHOT provides free transit within alternative modes of

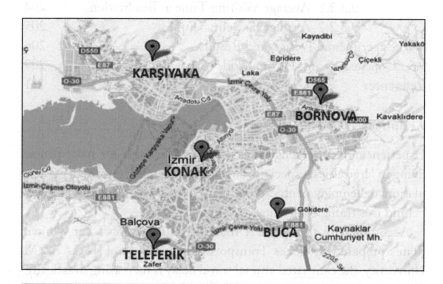

Figure 9.1   Bus operation districts of ESHOT.

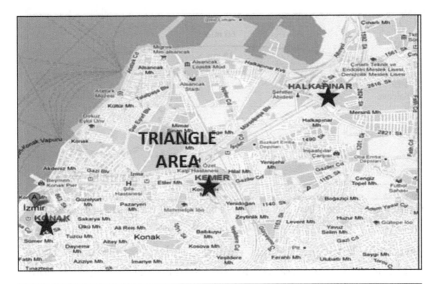

Figure 9.2 Triangle area in Izmir downtown.

public transportation for passengers when they use their smart tickets within 90 min of first use.

In Izmir city center, the area between Halkapınar Connection Centre, Kemer Connection Centre and Konak Connection Centre is known as the prestige location of Izmir. It has high concentration of businesses and shopping centers and historical structures. Figure 9.2 shows this area, which is also referred to as the *triangle area* throughout this study. As presented in Table 9.1, the high number of public buses that go in and out of the triangle area shows the potential congestion problem caused by these buses. Numbers show that approximately half of the ESHOT bus fleet actively performs in this area, and considering that the total surface area of Izmir province is 12,007 km^2 and city center's is approximately 816 km^2, the buses

Table 9.1 Facts of the Triangle Area

Number of bus lines in the triangle area	Total number of bus lines in Izmir	Percentage
90	317	0.29
Number of active buses in the triangle area	Total number of active buses in Izmir	Percentage
651	1408	0.46
Total bus exits and entrances in morning rush hours to the triangle area in a day (3 h)	Total bus exits and entrances to the triangle area in a day (18 h)	Percentage
2418	9480	0.26

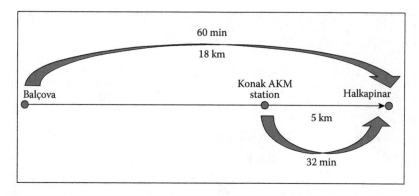

Figure 9.3   Example of time spent vs. distance traveled in the triangle area by buses.

have high density in the city center. Besides, comparison between total bus exits and entrances to the triangle area during rush hours and whole day shows that just 3 h morning rush hour period generates approximately one-fourth of the entire day's entrances and exits. Therefore, traffic congestion caused by public buses in the triangle area during rush hours results in higher congestion costs than any other period of time during the day.

As a result of traffic congestion, bus travel time increases in the triangle area. This means that the total transit time of passengers is too long compared to the distance of route in the triangle area. For instance, while the total traveling time of line 169 from Balçova to Halkapınar varies in the range of 54–60 min, it takes 26–32 min between Konak and Halkapınar, which means that almost half of the total time is spent in the triangle area while the triangle area's route length is approximately quarter of total route length of line 169 (Figure 9.3).

9.1.2 Objectives of the Study

In this study, a shuttle service system is proposed replacing the current bus routes and schedules in the triangle area of Izmir to solve the congestion problem. Figure 9.4 presents the proposed transfer hubs and the shuttle system for the *triangle area*. First, the simulation model of the current bus transportation system in the triangle area was developed, verified, and validated. Then, the expected benefits of the proposed shuttle system were determined through an experimental design using the simulation model.

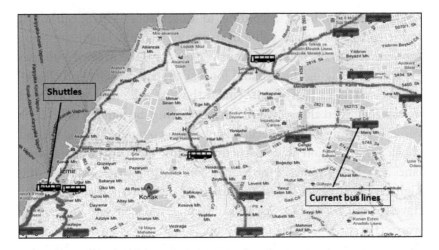

Figure 9.4   Proposed transfer hubs and shuttle system for the triangle area.

Eight different lines of shuttle services with different routes were planned to operate in the triangle area. Routes were based on work done jointly with ESHOT's transportation planning department. Some of the bus stations that were not frequently used in the current system were eliminated while determining shuttle services' routes. Expected outcomes of the proposed shuttle system were as follows:

- Decreasing average traveling and waiting time for passengers in the triangle area
- Decreasing the number of buses traveling in the triangle area
- Minimizing the simultaneous arrival of buses to bus stations (trailing)
- Decreasing total CO_2 emission

9.2 Literature Review

Many of the metropolitan cities suffer from higher traffic congestion in city centers. The factors behind this problem are numerous. The high number of business and entertainment centers can be counted as the first reason for the crowdedness. A great number of people travel into this region during the same few hours each morning and evening, called as *peak periods*; therefore, roads and public transportation systems do not have enough capacity for simultaneous arrival/exit of everyone who wants to use them (Downs, 2004).

On the other hand, according to Rosenbloom (1978), although traffic congestion is inevitable in metropolitan cities, there are some ways at least to decrease its intensity. These can be divided into two groups, as changing the demand for road system capacity and changing the system capacity itself. Under the first title, reorienting travel to less-congested alternative routes or reducing the number of vehicles while increasing vehicle occupancy can be counted. For the second choice, the solution is constructing additional roadway or adding lanes to existing routes. However, according to Parry, even when the highway capacity is increased, increase in the growth of vehicle miles traveled will be higher. As a result, congestion has grown steadily worse (Parry, 2002). In his study, some statistics is given to support this statement by using the Department of Transportation database. As an example, while vehicle miles traveled in urban areas increased by 289% between 1960 and 1991, total road capacity in urban areas increased by only 75%.

Improvements to decrease congestion can be increasing frequency and operating hours, improving coordination among different modes, providing real-time information to customers (GIS), or designing services that serve for particular travel needs, such as express commuter buses, special event service, and various types of shuttle services according to Transportation Demand Management Encyclopedia (Victoria Transport Policy Institute, 2013).

Other improvement suggestions for public transportation can be viewed in Boll's (2008) thesis in detail under the title of physical priority to buses. Grade-separated right of way, median bus ways, and contra-flow lanes built on one-way streets are some of the examples implemented worldwide for physical priority to buses. A video enforcement system can be implemented to control adherence to the rules when these systems are implemented.

Another group of methods to improve public transportation can be through incentives of using different modes of public transportation. A good example of this kind of incentive is *the linked transport* from Izmir, Turkey, which gives the chance of free ride to customers within 90 min after the first ride (ESHOT General Directorate). Madrid has also provided incentives to promote public transportation. Intermodal exchange stations for connections between urban and suburban transportation modes were built in Madrid to promote public transportation within the city. In the study of Vassallo et al. (2012), effects of this implementation was analyzed in terms of users, public transportation operators, infrastructure managers, the government, the abutters, and other citizens.

However, there is no universal measurement to analyze the effectiveness of any suggestions that are given earlier since *congested traffic* is a relative term. "In common sense, the traffic of any given artery can be considered congested when it is moving at speeds below the artery's designed capacity because drivers are unable to go faster" (Downs, 2004). Based on this concept, Texas Transportation Institute and the Federal Highway Administration developed some measures of congestion that includes the *travel time index*. This index is calculated as a ratio of the total travel time during rush hours to the total travel time during nonrush hours for the same route.

Simulation is a powerful tool to analyze all improvement suggestions given earlier and reach such an index, since it is possible to study detailed relations that might be lost in analytical or numerical studies. "The reasons to use simulation in the field of traffic are the same as in all simulation: the difficulty in solving the problem analytically; the need to test, evaluate and demonstrate a proposed course of action before implementation; to make research (to learn) and to train people" (Pursula, 1999). Also, Boxill and Yu (2000) claim that because of stable and unstable states, chaotic and stochastic behaviors of traffic, simulation is a useful method.

Olstam and Tapani (2011) define each step of developing a traffic simulation in detail. They state that firstly, aim and scope of the study should be determined before collecting necessary data. Following these steps, the simulation model can be constructed; however, it needs also verification, calibration, and validation. Thereafter, it is ensured that the model represents reality in a reasonable way, alternative scenarios should be tried, and each of them should be analyzed. They asserted that the final step should be documentation. According to Balci (1990), representation of reality in a reasonable way does not mean an absolute accuracy. He claims in his study that while in some cases 60% level of confidence is enough for the aim of the study, others can require 90% level of confidence.

Boxill and Yu (2000) categorize traffic simulation into three fields that are microscopic, macroscopic, and mesoscopic. While microscopic models focus on behaviors of individual vehicles such as speed and location or characteristics of drivers, macroscopic models aim to evaluate traffic flow or density as a continuum (Oner, 2004). On the other hand, a mesoscopic model integrates characteristics of these two approaches by considering individual vehicles' behavior and also general traffic flow.

9.3 Solution Method

In this section, the data collection and analysis and the developed simulation model are explained here.

9.3.1 Data Collection and Analysis

The main sources of data used in this study are the ESHOT smart ticketing system, GPS bus tracking database, and ESHOT transportation planning database.

Bus lines that perform in the triangle area were determined using ArcGIS, which is a complete system for designing and managing solutions through the application of geographic knowledge. Ninety bus lines that go through the triangle and all stations in the triangle area were determined. Entrance and exit points within the triangle area and all stations in the triangle area were identified for the 90 bus lines. Bus routes in the triangle area were identified from ESHOT's website. Distances between bus stations were retrieved from ESHOT's database in km.

ESHOT's smart ticket database keeps the information (date, time, direction) of each boarding passenger for each bus line. The October 2011 data were selected to be used since it is the most crowded month of that year. Rush hours were selected since a minor improvement in rush hour would definitely improve nonrush hours. Data for morning (06:00–09:00) and evening rush (16:00–20:00) hours were used. Three- and four-hour datasets were used instead of hourly datasets. The Kolmogorov–Smirnov two-sample test showed no statistically significant difference between using either way.

The travel time between consecutive stations on a line for each of the buses was fitted into a distribution using ARENA Input Analyzer.

The travel times between the bus stations can be generated in the simulation model using these distributions.

By using smart ticketing system database, arrival of each bus line to each station and the number of passengers that were accumulated between two consecutive arrivals of the buses were recorded. Assuming a steady flow of passengers to the stations, dividing time difference between two consecutive arrivals by the number of accumulated passengers shows the interarrival time of passengers. Interarrival times of passengers were calculated using Excel macro, and input analyzer was used for generating passenger interarrival time distributions. Data collection was made for every bus stop in the triangle area for each bus line. Data of passengers dropped off at each bus stop were provided by ESHOT, which was based on an estimate.

The smart ticketing system database was also used to identify the number of passengers for each line's first bus stop, namely, the Konak, Kemer, and Halkapınar bus stations.

9.3.2 Simulation Model

ARENA simulation software by Rockwell Automation was used to simulate both current system and proposed shuttle system. Although there were 90 bus lines that run through the triangle area, all of them were not included in the simulation model. Instead, these bus lines were grouped based on their entrance points to the triangle area, and then all three groups were scaled down to one-third. The resulting scaled-down bus lines were as follows:

- *Konak entrance point*: Lines 8, 12, 169, 300, 554, 42, 44, 45, 46, and 269
- *Kemer entrance point*: Lines 37, 38, 39, 42.44, 45, and 46
- *Halkapınar entrance point*: Lines 131, 140, 147, 148, 63, 576, 986, 886, 79, 70, and 169

Performance parameters of these models are representatives of all 90 bus lines.

The model starts by creating the passengers and the buses. When the bus arrives to station, first passengers are dropped off and then the

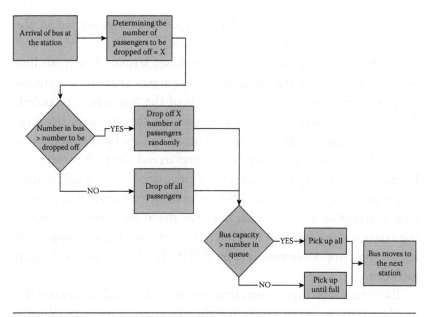

Figure 9.5   Flowchart of the simulation model for bus station events.

passengers waiting for the particular bus line are picked up. Finally, the bus moves to the next station.

Because of capacity restrictions, picking up and dropping of processes require some additional decision blocks in the model. These can be seen in detail in Figure 9.5, which shows the main logic of bus station events.

In this study, two different performance measures are given. These are as follows:

- Average waiting time of passenger at bus stations within the triangle
- Average traveling time of passenger in the triangle

9.3.2.1 Average Waiting Time at Bus Station   After passengers are created according to identified passenger interarrival time distributions by CREATE block, they arrive in stations' queues by using HOLD block in ARENA model. Passengers arrive in stations at different times, and when they get into the buses, average waiting time is calculated. Logic of this calculation is explained as follows:

Average waiting time of passengers at each station

$$
= \frac{\sum_{i=1}^{n} \left(T_{\text{now}} \left(\text{arrival of bus to the station} \right) - T_{\text{now}} \left(\text{arrival of the passenger } i \text{ to the station} \right) \right)}{\text{Total number of passengers gets into bus at station } i(n)}
$$

(9.1)

Then, average waiting time of passengers for each line was calculated as follows:

Average waiting time of passengers for each line

$$
= \frac{\sum_{i=1}^{n} \left(\text{Average waiting time of passengers at station } i * \text{Number of passengers at station } i \right)}{\sum_{i=1}^{n} \left(\text{Number of passengers at station } i \right)}
$$

(9.2)

9.3.2.2 Average Traveling Time   Traveling time defines the duration of passenger between getting into bus and getting out of bus. It is calculated as follows:

Average traveling time

$$
= \frac{\sum_{i=1}^{n} \sum_{j>i}^{n} \left(\text{Traveling time between stations } i \text{ and } j * \text{Number of passengers travel between stations } i \text{ and } j \right)}{\sum_{i=1}^{n} \sum_{j>i}^{n} \left(\text{Number of passengers travel between stations } i \text{ and } j \right)}
$$

(9.3)

The simulation model developed in ARENA was verified by comparing the number of buses created and the number of passengers picked up and dropped off at the bus stations with the data obtained from ESHOT for the current system.

9.4 Experimental Design and Results

After constructing the simulation model for shuttle services system, three parameters are selected to analyze the proposed system under different conditions. The three parameters are the expected percentage of passenger transfers to different transportation modes, frequency of shuttle services during rush hours, and expected % decrease in travel time due to the reduced number of buses in the triangle area. Each design parameter having three levels resulted in a total of 27 scenarios. Table 9.2 shows the experimental design parameters and their levels used in this study. Each scenario is run with 30 replications, and results from the scenarios are recorded.

The first parameter is the expected percentage of passenger transfers to different transportation modes such as ferry and subway. This transfer takes place when the passengers arrive at the shuttle service transfer hubs at the entrances of the triangle area. Three different levels for these parameters are selected as 10%, 15%, and 20%. These levels are estimated based on the passenger information from the smart ticketing system for different modes of transportation by ESHOT Transportation Planning Department. Different frequencies of the shuttle buses are selected as the second parameter and determined for each hour of the morning rush hours. First level of the shuttle bus frequencies is assigning shuttle buses in every 5 min between 6:00 and 7:00, in every 2 min between 7:00 and 08:00, and in every 4 min between 8:00 and 9:00. Other levels are scheduling shuttles in every 10, 5, 6 and 6, 4, 5 min between 6:00 and 7:00, 7:00 and 08:00, and 08:00 and 09:00, respectively. The frequency of the shuttle bus services varies during the 3 h period (morning rush hours) based on smart ticketing system data of the triangle area bus stations.

Table 9.2 Experimental Design Parameters and Their Levels

	PARAMETERS		
	EXPECTED PERCENTAGE OF PASSENGERS' TRANSFER TO DIFFERENT TRANSPORTATION MODES	DIFFERENT SCHEDULES FOR SHUTTLES BETWEEN 06:00 AND 07:00, 07:00 AND 08:00, AND 08:00 AND 09:00 (MIN)	PERCENTAGE OF EXPECTED DECREASE IN TIME PASS BETWEEN TWO CONSECUTIVE STATIONS
Levels	10	5, 2, 4	2
	15	6, 4, 5	4
	20	10, 5, 6	6

Third parameter is the expected decrease in bus travel time in the triangle area. Proposed shuttle service system decreases the number of buses used in the triangle area. Therefore, it is expected that the traffic congestion will be reduced, which might be reflected by decreased bus travel times. This reduction is shown in ARENA model by reducing the time between stations. Three levels for this parameter are 2%, 4%, and 6% reduction in bus travel time.

Each of eight shuttle lines are compared with current active buses on those routes and scenarios in terms of average passenger traveling time in the triangle area and average waiting time of passengers at bus stations.

Average travel time in the triangle and average waiting time at the bus stations for the current system are compared with the best- and worst-case scenarios of each shuttle line of the proposed system in Tables 9.3 and 9.4, respectively. In these tables, buses are grouped according to their routes and matched with shuttle system's lines. It is observed that even with the worst scenarios, almost all shuttle lines outperformed the current system.

The individual effects of the experimental design parameters, which make up the scenarios, can be better observed by plotting all of the scenarios and their improvements in two performance measures at the

Table 9.3 Comparison of Average Traveling Time of Passengers in the Triangle Area for the Current System with the Best- and Worst-Case Scenarios of the Proposed Shuttle System

	CURRENT	BEST		WORST	
		BUS GROUPS			
SHUTTLE LINE	BUS GROUP AVERAGE TRAVELING TIME OF PASSENGERS (S)	SHUTTLE AVERAGE TRAVELING TIME OF PASSENGERS (S)	% IMPROVEMENT OF BEST	SHUTTLE AVERAGE TRAVELING TIME OF PASSENGERS (S)	% IMPROVEMENT OF WORST
1	724.7	482.19	33.46	555.38	23.36
2	764.691	664.17	8.35	720.72	0.55
3	572.08	511.11	29.47	673.05	7.13
4	393.2	314.5	56.60	344.69	52.44
5	613.41	358.37	50.55	427.87	40.96
6	585.23	513.8	29.10	596.91	17.63
7	753.08	509.11	29.75	671.05	7.40
8	603.88	394.16	45.61	414.59	42.79

Table 9.4 Comparison of Average Waiting Time of Passengers at the Bus Stations for the Current System with the Best- and Worst-Case Scenarios of the Proposed Shuttle System

	CURRENT	BEST		WORST	
		BUS GROUPS			
SHUTTLE LINE	BUS GROUP AVERAGE WAITING TIME OF PASSENGERS (S)	SHUTTLE AVERAGE WAITING TIME OF PASSENGERS (S)	% IMPROVEMENT OF BEST	SHUTTLE AVERAGE WAITING TIME OF PASSENGERS (S)	% IMPROVEMENT OF WORST
1	430.14	152.09	79.01	312.74	56.85
2	469.85	145.34	79.94	956.13	−31.93
3	339.94	106.75	85.27	366.24	49.46
4	343.81	107.36	85.19	234.52	67.64
5	400.28	135.55	81.30	257.48	64.47
6	552.26	136.81	81.12	655.6	9.53
7	403.94	103.75	85.68	363.24	49.88
8	326.81	112.69	84.45	250.23	65.47

same time. Figure 9.6 shows the percentage improvements in average waiting time and average travel time for all 27 scenarios for shuttle line 1. Percentage improvements in Figure 9.6 clearly show three clusters. Upon closer examination, we found out that the clusters are almost perfectly formed based on the level of the second parameter,

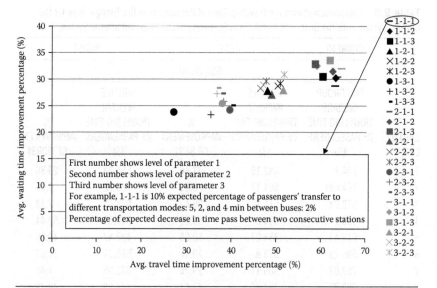

Figure 9.6 Shuttle line 1 travel time and station waiting time % improvements.

which is the shuttle bus frequency during the morning rush hour. Although the other two parameters also do affect the performance measures, bus frequency schedules seem to have the highest impact. Plots for the other shuttle lines show a similar trend.

With the reduction of the number of buses in the triangle area, CO_2 emission is also expected to decrease. In Table 9.5, results of CO_2 emission of current bus lines, which has the same routes with shuttle lines, are given. As it can be seen in Table 9.5, CO_2 emission tones/year can be approximately reduced to half even with the worst-case scenario. According to the UK Department for Environment, Food, and Rural Affairs (2012 DEFRA Database) database, the average CO_2 emission for local buses is 0.11195 kg/km. The total distance traveled within the triangle area is calculated, and the expected reduction for the CO_2 emission is estimated.

9.5 Conclusions and Future Work

Traffic congestion and its results are significant for many metropolitan cities around the world. Izmir is no exception. In this study, the focus was traffic congestion in Izmir city center due to the large number of public buses, especially during rush hours. An alternative shuttle system was proposed, which prevents the entrance of large public buses into the city center, called as the *triangle area.* In the proposed system, passengers transfer to either newly designed shuttle buses or to the alternative modes of public transportation system while traveling through the triangle.

Average passenger traveling times and average passenger waiting times at the bus stations were determined as performance indicators to compare current system with the proposed system. For each shuttle line, 27 different scenarios were generated based on three design parameters, and results were compared with current bus lines that give service on the same routes. Results of these scenarios showed significant improvements in performance measures, even for the worst-case scenarios. As an added benefit, the proposed shuttle system also had less CO_2 emissions due to the reduced number of buses in the triangle area.

Although, extra transfers made by passengers seem counterintuitive, our results prove that a well-designed system will improve

Table 9.5 Expected CO_2 Emission Reductions due to the Proposed Shuttle System in the Triangle Area

SHUTTLE LINE	CURRENT BUS GROUP AVERAGE CO_2 EMISSION (TONS/YEAR)	NUMBER OF SHUTTLES = 28 SHUTTLE AVERAGE CO_2 EMISSION (TONS/YEAR)	IMPROVEMENT (%)	NUMBER OF SHUTTLES = 37 SHUTTLE AVERAGE CO_2 EMISSION (TONS/YEAR)	IMPROVEMENT (%)	NUMBER OF SHUTTLES = 57 SHUTTLE AVERAGE CO_2 EMISSION (TONS/YEAR)	IMPROVEMENT (%)
1	0.05	0.02	60.0	0.03	40.0	0.04	20.0
2	0.06	0.02	66.7	0.03	50.0	0.05	16.7
3	0.06	0.02	66.7	0.03	50.0	0.04	33.3
4	0.03	0.01	66.7	0.01	66.7	0.02	33.3
5	0.05	0.02	60.0	0.03	40.0	0.04	20.0
6	0.06	0.02	66.7	0.03	50.0	0.05	16.7
7	0.06	0.02	66.7	0.03	50.0	0.04	33.3
8	0.03	0.01	66.7	0.01	66.7	0.02	33.3

BUS GROUPS

passenger experience and benefit the city as a whole. We believe, where the conditions are similar, simulation can be used efficiently to experiment with new system designs for public transportation systems.

Improvement on some of the estimation procedures is planned for future work. The estimates of the percentage of passengers transferring to other transportation modes and reduction in travel times in the triangle area due to the less number of buses were used in experimentation for evaluating alternative scenarios. In addition, although the passengers boarding at the stations were known due to the smart ticketing system, passenger destinations had to be estimated. More data on the passenger travel habits and traffic flow pattern in the triangle area with the use of a microsimulation package will improve the accuracy of the results of this study.

References

Balci, O. Guidelines for successful simulation studies. *Proceedings of the 1990 Winter Simulation Conference*, Piscataway, NJ. IEEE, New York, 1990, pp. 25–32.

Boll, C.M. Congestion protection for public transportation: Strategies and application to MBTA bus route 66. Thesis, Northeastern University, Boston, MA, 2008, pp. 8–14.

Boxill, S.A. and L. Yu. An evaluation of traffic simulation models for supporting ITS. Technical report. Center for Transportation Training and Research, Texas Southern University, Houston, TX, 2000.

Department of Environment, Food, and Rural Affairs. 2012 Guidelines to Defra/DECC's GHG conversion factors for company reporting: Methodology paper for emission factors. UK Department of Environment, Food, and Rural Affairs, London, U.K., July 2012.

Downs, A. *Still Stuck in Traffic: Coping with Peak-Hour Traffic Congestion.* The Brookings Institution, Washington, DC, 2004.

ESHOT General Directorate Webpage. Smart ticketing system operations. http://www.eshot.gov.tr/Faaliyet.aspx?MID=195 (accessed January 2014).

Olstam, J. and A. Tapani. A review of guidelines for applying traffic simulation to level-of-service analysis. *Sixth International Symposium on Highway Capacity and Quality of Service*, Stockholm, Sweden. Elsevier Ltd., Oxford, U.K., 2011, pp. 771–780.

Oner, E. A simulation approach to modeling traffic in construction zones. MSc thesis, Ohio University, Athens, OH, 2004.

Parry, I.W.H. Comparing the efficiency of alternative policies for reducing traffic congestion. *Journal of Public Economics* 85 (2002): 334.

Pursula, M. Simulation of traffic systems—An overview. *Journal of Geographic Information and Decision Analysis* 3 (1999): 1–8.

Rosenbloom, S. *Peak Period Traffic Congestion: A State of the Art Analysis and Evaluation of Effective Solutions.* Elsevier Scientific Publishing Company, Amsterdam, 1978, p. 169.

Schrank, D., B. Eisele, and T. Lomax. TTI's 2012 urban mobility report. Yearly mobility report. Texas A&M Transportation Institute, College Station, TX, 2012.

Vassallo, J.M., F. Di Ciommo, and A. Garcia. Intermodal exchange stations in the city of Madrid. *Transportation* (Springer Science+Business Media, LLC) 29 (2012): 975–995.

Victoria Transport Policy Institute. Public transit improvements. Prod. Victoria Transport Policy Institute, Victoria, British Columbia, Canada, August 28, 2013.

10

Relocation of the Power Transmission and Distribution Division of a Multinational Electronics and Electrical Engineering Company

MESUT KUMRU

Contents

10.1 Introduction

Facility layout and design are an important issue for any business entity's overall operations, in terms of both maximizing the effectiveness of production processes and meeting the employee needs and/or desires. Facility layout is defined by Weiss and Gershon (1993) as "the physical arrangement of everything needed for the product or service, including machines, personnel, raw materials, and finished goods. The criteria for a good layout necessarily relate to people (personnel and customers), materials (raw, finished, and in processes), machines, and their interactions."

Business owners need to consider many operational factors when building or renovating a facility for maximum layout effectiveness. These factors include the following: future expansion or possible changes of facility, land use, workflows, material movements, transportation and procurement needs, output requirements, ease of communication and support, employee morale and job satisfaction, promotional values, and safety. In order not to continuously redesign the facility, the facility layout problem should be handled very carefully. There are many goals in facility design such as keeping the material movement at a minimum level, avoiding bottlenecks, minimizing machine interventions, enhancing employee morale and security, and providing flexibility.

There are three basic types of layouts: product, process, and fixed position. Three hybrid types of layouts are also used: cellular, flexible manufacturing systems, and mixed-model assembly lines. Essentially, two distinct types of layout (product and process) are widely implemented. Product layout mainly affects the assembly line arrangement and is very much concerned with the products produced. Process layout, on the other hand, is established according to the production processes that are used to generate the products. Product layout is principally applied to high-volume repetitive operations, while process layout is applied to low-volume make-to-order operations.

Carefully planning the layout of a facility can have significant long-term benefits for the company's manufacturing and distribution activities. Creating a sustainable growth plan is an essential key to develop this plan. Many issues (production process routings and flows, material handling methods and equipment requirements, product mix and volumes, etc.) must be considered while developing this plan.

Basic purpose of layout is to ensure a smooth flow of work, material, and information through the system. However, a lot of objectives are considered to achieve that: minimization of material handling costs; efficient utilization of space and labor; elimination of bottle-necks; facilitation of communication and interaction between work-ers, between workers and their supervisors, and/or between workers and customers; reduction of manufacturing cycle time and customer service time; elimination of wasted or redundant movement; facilita-tion of the entry and exit; placement of material, products, and people; incorporation of safety and security measures; promotion of product and service quality; encouragement of proper maintenance activities; providing a visual control of operations or activities; and providing flexibility to adapt to changing conditions.

In designing process layouts, the most significant objective is to minimize material handling costs. This implies that departments that incur the most interdepartmental movement should be located closest to one another. For this purpose, two main approaches are widely used to design layouts, which are algorithmic and procedural approaches (Yang et al., 2000). Algorithmic approaches consider only quantita-tive factors and do not consider any qualitative factors, whereas proce-dural approaches can use both. Algorithmic approaches can efficiently generate alternative layout designs with often oversimplified objec-tives (Yang and Hung, 2007). They can be computationally complex and prohibitive. That is why systematic layout planning (SLP) was adopted in industries as a viable approach in the past few decades (Han et al., 2012). Therefore, a procedural layout design approach— SLP—is preferred in this chapter to solve the facility relocation problem of an electronics and electrical company. Furthermore, the performance of the preferred method is compared to those of Nadler's ideal systematic approach (another procedural approach) and delta-hedron (a graph theoretic-based heuristic algorithm), by use of linear weighting in factor analysis. After giving the facility layout problem definition and literature survey results along with the introduction of the techniques used in the study, details of the application are given in the following sections.

10.2 Facility Layout Problem

The placement of facilities on the plant site is often known as *facility layout problem*. This activity has a significant influence on manufacturing costs, operation processes, lead times, and productivity. A suitable placement of facilities contributes to the overall efficiency of the plant and reduces the operating expenses up to 50% (Tompkins et al., 1996). Simulation studies are usually carried out to measure the benefits and performance of given layouts (Aleisa and Lin, 2005). Since layout problems are known to be complex and generally NP-hard (Garey and Johnson, 1979), numerous research studies were conducted in this area during the past decades.

As researchers have taken into consideration various ideas in their studies, they could not agree on a standard and exact definition of layout problems. A facility layout is an arrangement of everything needed for the production of goods or delivery of services. A facility is an entity that facilitates the performance of any job. It may be a machine tool, a work center, a manufacturing cell, a machine shop, a department, a warehouse, etc. (Heragu, 1997). Koopmans and Beckmann (1957) defined the facility layout problem as a common industrial problem where the objective is to configure facilities in a way to minimize the cost of transporting materials between them. Azadivar and Wang (2000) reported that the facility layout problem is the determination of the relative locations for a given number of facilities and allocation of the available space among those facilities. According to Lee and Lee (2002), the facility layout problem consists in arranging unequal-area facilities of different sizes within a given total space, which can be bounded to the length or width of site area so as to minimize the total material handling and slack area costs. Shayan and Chittilappilly (2004) defined the facility layout problem as an optimization problem that tries to make layouts more efficient by considering various interactions among facilities and material handling systems while designing layouts.

Drira et al. (2007) stated that the problems addressed in research works differ depending on such factors as follows:

Workshop characteristics impacting the layout: Products variety and volume, facility shapes and dimensions, material handling systems, multifloor layout, backtracking and bypassing, and pickup and drop-off locations.

Static versus dynamic layout problems (formulation of layout problems): Discrete formulation, continual formulation, fuzzy formulation, multiobjective layout problems, and simultaneous solving of different problems.

Resolution approaches: Exact approaches and approximated approaches.

Above all, recent papers rest on complex and realistic features of the manufacturing systems studied. Facility layout is taken into consideration together with typical parameters such as pickup/deposit points, corridors, and complex geometric constraints, when formulating the layout design problem. A lot of research contains restrictive assumptions that are not adapted to the complexity of many manufacturing system facilities. This is an outdated approach and certainly an important issue that should be considered (Benjaafar et al., 2002). However, research is still needed. Designing a plant using a third dimension as a recent approach necessitates more research, such as to select and optimize resources related to the vertical transportation of parts between different floors.

Researchers have preferred mostly to deal with static layout problems rather than dynamic ones. However, considering the changing conditions of operation systems, it is clear that the static approaches are unable to follow up these changes. The dynamic approaches have been developed against these changing business conditions in the future and are sometimes seen as good alternatives. Also, fuzzy methods may offer possibilities to assess uncertainty. Meanwhile, as already noted by Benjaafar et al. (2002), research is still needed for suggesting or improving methods to design (1) robust and adaptive layouts, (2) sensitivity measures and analysis of layouts, and (3) stochastic models used to evaluate solutions.

When methods used in the solution of layout problems are concerned, it is seen that the metaheuristic methods have been widely used in facility layout studies dealing with problems in a larger size and taking into account constraints in a more realistic way. Evolutionary algorithms seem to be among the most popular approaches. Solution methods are also hybridized (integrated) to solve complex problems or to develop more realistic solutions. The studies based on artificial intelligence are now rarely published.

On the other hand, due to the difficulty in solving any problem without the use of expert systems, hybrid methods, capable of optimizing the layout, are likely to be still needed while taking into account the available expert knowledge.

Most of the published research has focused on the determination of plant layout. However, in practice, this problem is often addressed with other design issues like the selection of production or transportation source, the design of cells, and the determination of capacity resources. These problems are generally dependent on each other, for example, selection of a material handling conveyor as a means of transportation induces the selection criteria of automated guided vehicles. Therefore, during the plant design, research is needed to bring solutions to a variety of problems addressed simultaneously rather than sequentially. Such studies are promising in solving problems toward development and improvement of plant layout. This approach will indeed direct the researchers to focus on workshop design problems rather than being concentrated only on facility location problems.

10.3 Literature Search

Facility layout design approaches in the literature are commonly categorized as algorithmic and procedural approaches (Yang et al., 2000). Algorithmic approaches can efficiently generate alternative layout designs with often oversimplified objectives (Yang and Hung, 2007). In these approaches, quantitative use of material handling distances and loads are used to develop layout alternative with minimum total material handling cost. Since these approaches take the flow distance, either measured in Euclidean or rectilinear distance, which may not represent the physical flow distance, they simplify both design constraints and objectives in order to achieve surrogate objective function for attaining the solution. When qualitative design criteria are concerned, these approaches cause lack of functionality and credence for a quality solution. The shortcoming of qualitative approaches comes out when all qualitative factors are aggregated into one criterion. These approaches can generate better results when commercial software is available. The basic limitation of these approaches is that they consider only quantitative factors and do not consider any qualitative factors. Some additional approaches in this category used the flow distance as the surrogate function to solve the layout design problem by utilizing mixed-integer programming formulation (Heragu and Kusiak, 1991; Peters and Yang, 1997; Yang et al., 2005; Chan et al., 2006), but they were often computationally prohibitive. Heuristics, metaheuristics, neural network, and fuzzy logic were also utilized in generating layout alternatives as well as exact procedures (Singh and Sharma, 2006). The majority of the existing literature reports on algorithmic approaches (Heragu, 1997).

On the other hand, procedural approaches can incorporate both qualitative and quantitative objectives in the design process, which is divided into several steps that are then solved sequentially (Han et al., 2012). These approaches rely on experts' experience (Yang et al., 2000). An effective and most famous method in this category is known as SLP procedure (Muther, 1973). SLP is widely used among enterprises and the academic world. The practical applications in a traditional SLP require intricate steps that can lead to lack of stability in results, if not applied properly. Since algorithmic approach requires for advanced training in mathematical modeling techniques, SLP was adopted in industries as a viable approach in the past few decades (Han et al., 2012). Chien (2004) proposed new concepts and several algorithms to modify procedures and enhance practicality in traditional SLP. In order to solve a factory layout design problem, Yang et al. (2000) applied the SLP as infrastructure and then the AHP for evaluating the design alternatives. Considering the hygienic factors, Van Donk and Gaalman (2004) utilized SLP in planning the layout of food industry. Based on SLP and AHP, a cellular manufacturing layout design was applied to an electronic manufacturing service plant (Nagapak and Phruksaphanrat, 2011). Mu-jing and Gen-gui (2005) combined SLP and the genetic algorithm to solve facility layout problem. As different from the earlier studies, Han et al. (2011) proposed parametric layout design for a flexible manufacturing system.

The SLP method proposed in this study is a practical approach for new layout designs that do not require deep mathematical knowledge.

Its performance is compared in this work to those of graph theoretic-based deltahedron heuristic and procedure-based Nadler's approaches. All three approaches are shortly described in the next section.

10.4 SLP, Deltahedron, and Nadler's Approaches

The SLP procedure (Muther, 1973) uses sequential steps in solving the layout problem. It is based on the input data and an understanding of the roles and relationships between activities. In the first step, a material flow analysis (from–to chart) and an activity relationship analysis (activity relationship chart) are performed. From these analyses, a relationship diagram is developed to be used as the foundation of the procedure. The next two steps involve the determination of the amount of space to be assigned to each activity, and the allocation of the total space to the departments, considering the relationship diagram. The criterion (objective measurement) for the positioning of departments is the department adjacency or some other user-defined metrics. Irrelevant adjacencies are reflected by zero scores, while relevant adjacencies are demonstrated by relationship/adjacency scores determined as the number of interdepartmental material flows. Based on modifying considerations and practical limitations, a number of layout alternatives are developed and evaluated. The preferred alternative is then recommended.

Deltahedron heuristic was developed by Foulds and Robinson (1978) to construct a maximal planar adjacency graph by a sequence of *insertions* of a new vertex and three edges into a triangular face. It starts with a complete graph on four vertices (K_4). Input requirements are the initial K_4 and the *insertion order* in which the vertices will be processed. Each vertex is successively inserted into the face of the triangulation that results in the largest increase in edge score (Giffin et al., 1995). The objective is maximizing relationship (adjacency) scores. Once the adjacency graph has been obtained, a corresponding block layout is then constructed.

Nadler's ideal system approach (Nadler, 1961) is one of the procedural approaches to facility layout planning and is based on the following hierarchical steps: (1) aim for the *theoretical ideal system*, (2) conceptualize the *ultimate ideal system*, (3) design the *technologically workable ideal system*, and (4) install the *recommended system*.

10.5 Case Study

10.5.1 Company

The case company is a global powerhouse in electronics and electrical engineering, operating in manufacturing, energy, and health-care sectors. It has operations in almost 190 countries including Turkey and owns approximately 285 production and manufacturing facilities with nearly 470,000 employees around the world. The company has operation facilities in Istanbul as well and is already producing various products in its Kartal plant. The range of products in Kartal plant varies in 14 main categories from middle voltage panels to protection systems. Around 2500 employees are hired in Kartal plant.

10.5.2 Problem

Among its production divisions, the power transmission and distribution (PTD) division has achieved the highest growth rates in the past years. Since it has reached to its full capacity, an urgent facility revision was needed for the division. The assembly and the tooling areas, as well as the preproduction, preassembly, and storage areas were not sufficient to meet new demands. Factory management decided to conduct a research on relocating the PTD division in another separate area. For this purpose, a project team was established to carry out the project.

The new facility area was taking place in Gebze, a district of Kocaeli city, and was more than two times larger than the existing facility area in Kartal. The current facility area was about 9,000 m², while the new area was almost 20,000 m². Because of this fact, an entirely different and current layout-centric new facility placement was necessary for the plant. The new facility would also include a new department called voltage circuit breaker (VCB). The products of VCB department were formerly being outsourced, but due to quality and transportation cost reasons, the company decided to establish this department within the company. One of the most important problems was the placement of the quality control (QC) department as it was outside the main area. In addition, the dispatching department was experiencing a shortage of place. Though all of the departments (except VCB) were taking place under the same roof in the current facility, the available site was not sufficient to meet the increasing customer demand due to unexpected high growth rates. Therefore, instead of redesigning the existing layout in regard to the needs of QC, VCB, and dispatching departments, they were transferred to a particular production area outside the plant as an urgent solution. In this way, additional places of production adjacent to the main location were provided. However, in the meantime, material handling costs increased and also waste of time became an important matter because of different production places. Thus, a new settlement (relocation) of the division was considered unavoidable.

According to the research and forecasts, the current production capacity was almost full, and the existing production area would not be adequate in the coming 3 years. Hence, the relocation of the production site has become a matter of priority for the company. This issue was addressed in a project that would include various alternative layouts for the new area. The problem was not only changing the current layout, but it was more than that. A relocation plan of the production facility would be established with a brand-new design. After all alternative layouts were prepared, they would be compared in terms of effectiveness scores, and the layout best to solve the problem would be chosen.

10.5.3 Methodology

While solving the relocation problem, the machine and equipment costs were ignored; the material flow and the departmental needs were taken into account. The figures about the sizes of areas required for the new project were attained from the previous research projects. In designing the layout plans, SLP technique was used. The SLP technique was based on the quantitative data as well as the identification of the roles and relationships between production activities. A material flow analysis (from–to chart) and an activity relationship analysis (activity relationship chart) were constructed. The relationship diagram positioned activities spatially. Proximities were typically used to reflect the relationship between pairs of activities. The next step involved the determination of the amount of space to be assigned to each activity. The SLP procedure, which depends on the relationship scores between the departments, was used to develop first a block layout and then a detailed layout regarding each depart-ment. If two departments' borderlines are touching each other, the score was calculated; otherwise, it was considered "0". All the inter-departmental relationships were quantified according to this prin-ciple. After preparing the relationship diagram, three alternative layout solutions (rectangular and L shapes) were developed. Two more layout solutions were also generated by using graph theoretic-based deltahedron heuristic and Nadler's ideal system approach, for performance comparison of the techniques. All of the alternative layout solutions were compared by use of linear weighting on the basis of the criteria of relationship/adjacency, waiting time, flexibil-ity, safety, and ease of supervision. Three alternative layouts were developed. Depending on the results, the performances of the tech-niques were evaluated comparatively, and the most effective solution was recommended.

10.5.4 Relocating the PTD Division

10.5.4.1 Production Capacity, Departments, and Their Abbreviations in PTD Division   In the middle voltage product range of the company, there are three main kinds of products: BT panels, BK panels, and Simoprime. BT panels have also three subgroups identified as 8BT1, 8BT2, and 8BT3. Hence, the total number of product groups is five (8BT1, 8BT2, 8BT3, 8BK20, and Simoprime). Production capacities as to the groups of products are given in Table 10.1.

In PTD division, 13 departments are operating. The names of those departments and their abbreviations used in the proceeding sections are given in line with the process flow in Table 10.2.

Table 10.1   Production Capacities

NO.	PRODUCT	PRODUCT QUANTITY/3 YEAR	MONTHLY AVERAGE	PIE%
1	8BT1	74	3	2
2	8BT2	954	27	16
3	8BT3	160	5	3
4	8BK20	1887	53	32
5	Simoprime	2841	79	47

Table 10.2   Departments and Their Abbreviations In Line with the Process Flow

NO.	DEPARTMENT	ABBREVIATION	NO.	DEPARTMENT	ABBREVIATION
1	Warehouse	WH	8	Voltage circuit breaker	VCB
2	Trumatic (Punching)	TR	9	Truck assembly	TA
3	Bending	B	10	Main assembly	MA
4	Welding	W	11	Quality control	QC
5	Painting	P	12	Crash area	CA
6	Logistic center	LC	13	Dispatch	D
7	Low-volt panel assemb.	LV			

Table 10.3   Minimum Space Requirements of the Departments

NO.	DEPT.	SPACE (M²)	NO.	DEPT.	SPACE (M²)	NO.	DEPT.	SPACE (M²)
1	WH	900	6	LC	1300	11	QC	1,700
2	TR	700	7	LV	1700	12	CA	600
3	B	800	8	VCB	700	13	D	1,500
4	W	700	9	TA	700	Total		16,000
5	P	1600	10	MA	3000			

The minimum space requirement of each department in PTD division is determined as given in Table 10.3.

The activity relationships represented by codes indicating which activities are in relation to each other are given in Table 10.4.

The annual number of material flows between departments is called as score. The closeness (adjacency) rating as to the scores is determined with respect to the maximum score of 958 as given in Table 10.5.

Table 10.4   Activity Relationships

	WH	TR	B	W	P	LC	LV	VCB	TA	MA	QC	CA	D
A	TR	—	—	—	MA	VCB	MA	TA	—	—	D	—	—
E	—	—	—	P	—	—	—	MA	MA	—	—	—	—
I	—	—	W, P	—	—	—	—	—	—	QC	—	—	—
O	—	B	—	MA	—	LV	—	—	—	—	—	—	—
U	—	—	—	—	TA	TA, MA	—	—	—	—	CA	—	—

Table 10.5   Closeness Rating as to the Scores

A (ABSOLUTELY NECESSARY)	E (ESPECIALLY IMPORTANT)	I (IMPORTANT)	O (ORDINARY)	U (UNIMPORTANT)
766.4–958	574.8–766.4	383.2–574.8	191.6–383.2	0–191.6

From \ To	WH	TR	B	W	P	LC	LV	VCB	TA	MA	QC	CA	D
WH		786 (A)	0 (U)	0 (U)	0 (U)	0 (U)	0 (U)	0 (U)	0 (U)	0 (U)	0 (U)	0 (U)	0 (U)
TR			294 (O)	0 (U)	0 (U)	0 (U)	0 (U)	0 (U)	0 (U)	0 (U)	0 (U)	0 (U)	0 (U)
B				512 (I)	480 (I)	0 (U)	0 (U)	0 (U)	0 (U)	0 (U)	0 (U)	0 (U)	0 (U)
W					723 (E)	0 (U)	0 (U)	0 (U)	0 (U)	308 (O)	0 (U)	0 (U)	0 (U)
P						0 (U)	0 (U)	0 (U)	69 (U)	958 (A)	0 (U)	0 (U)	0 (U)
LC							303 (O)	881 (A)	182 (U)	146 (U)	0 (U)	0 (U)	0 (U)
LV								0 (U)	0 (U)	833 (A)	0 (U)	0 (U)	0 (U)
VCB									880 (A)	727 (E)	0 (U)	0 (U)	0 (U)
TA										622 (E)	0 (U)	0 (U)	0 (U)
MA											542 (I)	0 (U)	0 (U)
QC												36 (U)	940 (A)
CA													0
D													

Figure 10.1 From–to chart with closeness ratings.

The relationships between departments (from–to chart) are given in Figure 10.1, in terms of the corresponding scores and closeness ratings.

10.5.4.2 Analysis of the Existing Layout The current block layout of the PTD division is shown in Figure 10.2. The interdepartmental closeness ratings as to the scores and the total score attained for the existing layout are given in Table 10.6.

10.5.4.3 Alternative Layouts Based on the current layout input, five new alternative layouts were generated for the PTD division to be established in the new plant area in Gebze region. The first three layout alternatives were generated by the use of SLP approach, while

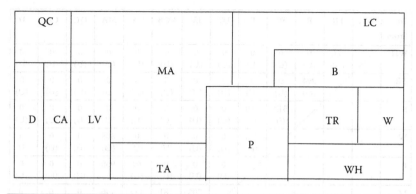

Figure 10.2 Existing block layout (Kartal).

Table 10.6 Relations and Scores for the Existing Layout

DEPARTMENTS	WEIGHTS	DEPARTMENTS	WEIGHTS	DEPARTMENTS	WEIGHTS
LV–MA	833	B–TR	294	CA–D	0
LV–TA	0	B–W	512	QC–LV	0
MA–LC	146	TR–W	0	QC–MA	542
MA–P	958	TR–WH	786	D–LV	0
MA–TA	622	W–WH	0	B–P	480
LC–P	0	QC–CA	36	TA–P	0
LC–B	0	QC–D	940	*Total*	*6149*

the fourth and the fifth alternative layouts were generated by the use of deltahedron and Nadler's approaches, respectively. Three of these layout alternatives are of rectangular shape, and the remaining two are of L shape. These layouts and their corresponding total scores are given in Figures 10.3 through 10.7 and Tables 10.7 through 10.11. Adjacency was adopted as the definition of closeness. If two activities had a common border, they were judged to be close; otherwise, they were not.

Detailed drawings of the current and alternative/candidate layouts were generated by the use of AutoCAD software. Alternative configurations with varying placements of departments were tested several times in order to find the best settlement.

10.5.5 Evaluation of Alternative Layouts

The alternative layouts developed in this study differed from each other with respect to the relationship scores. When only relationship

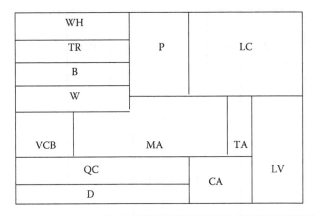

Figure 10.3   Alternative layout 1.

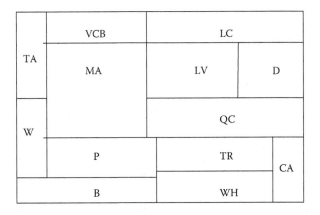

Figure 10.4   Alternative layout 2.

scores are considered, the second SLP alternative seems to provide the best solution with its highest score of 9531. The other alternatives take place in a descending order by score as Nadler's (8124), delta-hedron (7761), SLP1 (7559), and SLP3 (6539). The best solution is of rectangular type, but the second one is of L shape. Deltahedron took the third row with its block design. Though SLP has generated the best solution with its block diagram (SLP2), it has also generated the worst one with its L-shape diagram (SLP3). It seems that proce-dural approaches of SLP and Nadler's have provided good solutions by taking the first two rows in the list. The graph theoretic-based deltahedron approach could have taken only the third row. When

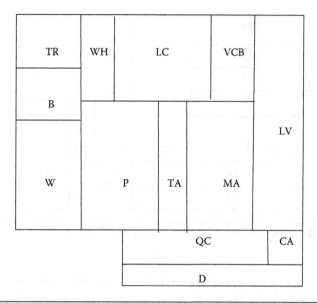

Figure 10.5 Alternative layout 3.

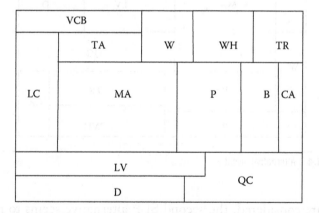

Figure 10.6 Deltahedron layout (layout 4).

the first three layout alternatives were regarded, the SLP method seemed superior to others.

Though the relationship score is an objective measurement for layout planning, it is not usually sufficient to select the proper layout and should be supported with some other criteria that are also influential in decision-making process. In our case, four more criteria were taken into consideration along with the relationship (adjacency) score in order to select the appropriate layout. These additional criteria were

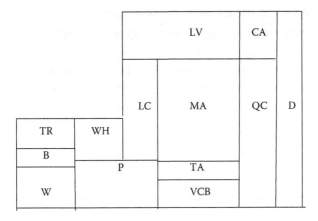

Figure 10.7   Nadler's layout (layout 5).

Table 10.7   Relations and Scores for Alternative Layout 1

DEPARTMENTS	WEIGHTS	DEPARTMENTS	WEIGHTS	DEPARTMENTS	WEIGHTS
WH–P	0	VCB–MA	727	LC–MA	146
WH–TR	786	QC–D	940	TA–MA	622
TR–B	294	QC–MA	542	TA–LV	0
TR–P	0	CA–LV	0	CA–TA	0
B–W	512	CA–QC	36	MA–P	958
B–P	480	CA–D	0	W–P	723
W–VCB	0	LC–P	0	MA–CA	0
W–MA	308	LC–TA	182	*Total*	*7559*
VCB–QC	0	LC–LV	303		

Table 10.8   Relations and Scores for Alternative Layout 2

DEPARTMENTS	WEIGHTS	DEPARTMENTS	WEIGHTS	DEPARTMENTS	WEIGHTS
TR–WH	786	LC–TA	0	MA–W	308
B–WH	0	LC–MA	0	LV–CA	0
B–W	512	P–MA	958	LV–QC	0
B–P	480	P–QC	0	QC–CA	36
W–P	723	TA–QC	0	QC–D	940
WH–LC	0	TA–MA	622	CA–D	0
WH–P	0	MA–VCB	727	TA–VCB	880
LC–LV	303	MA–LV	833	*Total*	*9531*
LC–VCB	881	MA–QC	542		

Table 10.9 Relations and Scores for Alternative Layout 3

DEPARTMENTS	WEIGHTS	DEPARTMENTS	WEIGHTS	DEPARTMENTS	WEIGHTS
TR–WH	786	LC–TA	182	MA–CA	0
B–WH	0	LC–MA	146	LV–CA	0
B–W	512	P–TA	69	LV–QC	0
B–P	480	P–QC	0	QC–CA	36
W–P	723	TA–QC	0	QC–D	0
WH–LC	0	TA–MA	622	CA–D	0
WH–P	0	MA–VCB	727	Total	6539
LC–P	0	MA–LV	833		
LC–VCB	881	MA–QC	542		

Table 10.10 Relations and Scores for Deltahedron Layout

DEPARTMENTS	WEIGHTS	DEPARTMENTS	WEIGHTS	DEPARTMENTS	WEIGHTS
TR–WH	786	LC–TA	182	MA–P	958
B–WH	294	LC–MA	146	LV–CA	0
B–TR	0	VCB–TA	880	LV–QC	0
B–P	480	P–QC	0	QC–CA	36
W–P	723	TA–QC	0	QC–D	940
WH–LC	0	TA–MA	622	CA–D	0
WH–P	0	MA–VCB	0	Total	7761
LC–P	0	MA–LV	833		
LC–VCB	881	MA–QC	0		

Table 10.11 Relations and Scores for Nadler's Layout

DEPARTMENTS	WEIGHTS	DEPARTMENTS	WEIGHTS	DEPARTMENTS	WEIGHTS
TR–WH	786	LC–MA	146	QC–CA	36
TR–B	294	P–LC	0	CA–D	0
B–WH	0	TA–VCB	880	QC–D	940
W–B	512	P–TA	69	QC–MA	542
B–P	480	VCB–P	0	TA–QC	0
W–P	723	MA–LV	833	VCB–QC	0
WH–LC	0	MA–TA	622	Total	8124
WH–P	0	MA–P	958		
LC–LV	303	LV–CA	0		

flexibility, safety, waiting time, and ease of supervision. Flexibility criterion in designing the facility layout was concerned with taking into account the changes over short and medium terms in the production process and manufacturing volumes. Safety criterion was concerned with safety level in the movement of materials and personnel workflow. Waiting time criterion referred the elapsed time during the workflow. Ease of supervision was related to the complexity of the workflow. *Factor analysis technique by linear weighting* was applied to the selection criteria that have impact on the facility layout decision. First, appropriate weights were assigned by the plant experts to each criterion on the relative importance of each. Later, alternative layouts were assessed one by one on the basis of those four criteria. Scores over 100 were assigned to each layout alternative with respect to the criteria identified (Table 10.12). While assigning scores to the criteria for each layout alternative, experts took into account several inputs, like minimum movement of people, material, and resources; space allocation and free space area; complexity and density of the layout; and interdepartmental disconnection distances. After normalization of the scores, total weight for each layout alternative was computed, and the one with the highest score was selected as the best alternative (Table 10.13).

Table 10.12 Scores of the Layout Alternatives as to the Criteria

METHODS	RELATIONSHIP SCORE	FLEXIBILITY	SAFETY	WAITING TIME	EASE OF SUP
SLP1	7559	20	50	60	40
SLP2	9531	40	75	80	60
SLP3	6539	70	40	40	80
Deltahedron	7761	50	60	70	50
Nadler's	8124	100	20	30	100

Table 10.13 Normalized Scores and Total Weights of the Layout Alternatives

METHODS	RELATIONSHIP SCORE (0.70)	FLEXIBILITY (0.15)	SAFETY (0.08)	WAITING TIME (0.05)	EASE OF SUP (0.02)	TOTAL WEIGHT
SLP1	0.19	0.07	0.20	0.21	0.12	0.1724
SLP2	0.24	0.14	0.31	0.29	0.18	0.2319
SLP3	0.16	0.25	0.16	0.14	0.25	0.1743
Deltahedron	0.20	0.18	0.25	0.25	0.15	0.2025
Nadler's	0.21	0.36	0.08	0.11	0.30	0.2189

When we look at Table 10.13, it is seen that the relationship score was given the highest weight of 0.70, while the other criteria took weights in between 0.15 and 0.02. This implies that the plant experts had selected the relationship criterion as the most important one affecting the layout decision. According to the total weights, the lay-out alternatives took place in a descending order by weight as SLP2 (0.2319), Nadler's (0.2189), deltahedron (0.2025), SLP3 (0.1743), and SLP1 (0.1724). Though this is quite consistent with the ordering of layouts with respect to relationship scores, only the SLP1 layout was negatively affected from flexibility and ease of supervision criteria when compared to other alternatives and took the last row instead of the forth row. Hence, SLP2 layout alternative has proved to be supe-rior to Nadler's and deltahedron alternatives and has been selected as the best one. If we think that the relationship score of the existing layout is 6149, then the relationship score of the layout selected is about 55% better than the existing layout with its relationship score of 9531. Even if it cannot be easily measured, we may expect a quite high overall efficiency by implementing the selected layout in the PTD division.

These five alternative layouts bring nearly optimal solutions. Theoretically, it is possible to find better solutions as well. But when we regard the applicability (building limitations, etc.) of those solu-tions, the proposed alternative layouts stand out feasible compared to the others.

10.6 Conclusion

The most important cause of high material handling costs is the lack of strategic facility planning. In an effective layout, material handling costs can be reduced by 10%–30%. This situation affects the cost of production directly.

This research was conducted in line with the expectations and needs of a multinational company to find an urgent appropriate solu-tion to the relocation problem of one of its production divisions, namely, PTD. The PTD division was incurring high amount of han-dling costs that necessitated an urgent layout revision. On the way to solve the problem, first, the needs of the division were defined later; the solution was reached easily by the use of SLP, Nadler's, and deltahedron methodologies, and the linear weighting for criteria comparisons. SLP is found to be the most effective and widely used alternative method in this kind of relocation problems. In our application, five alternative layouts were generated including the most effective one. Of course, several other methods could be used in such problems too. It is expected that the handling cost was reduced about 50% at PTD division after relocation.

Each layout design application is unique in nature, that is, there are different attributes associated with different applications; thus, the success of the present study has no guarantee for its applicability to other applications. Judicious use of a design method is advised in solving a specific application.

The project implementation given in this chapter focuses on manufacturing system applications. In fact, layout design problems exist in almost every type of system, such as manufacturing, hospitals, hotels, ports, and supermarkets. However, advances made in the specific areas of manufacturing may lead to positive influences on designing the layouts for other systems.

Finally, we should refer to commercial software tools available on the market developed for assisting in the design of manufacturing layout problems. Though these tools have been developed considering the manufacturing systems, they are limited in number. Therefore, additional software tools with generic solution approaches are needed in order to bring easy and quick solutions to every type of layout design problems.

References

Aleisa, E.E. and Lin, L. (2005). For effectiveness facilities planning: Layout optimization then simulation, or vice versa? *Proceedings of the 37th Winter Simulation Conference.* Orlando, FL, pp. 1381–1385.

Azadivar, F. and Wang, J. (2000). Facility layout optimization using simulation and genetic algorithms, *International Journal of Production Research*, 38(17), 4369–4383.

Benjaafar, S., Heragu, S.S., and Irani, S.A. (2002). Next generation factory layouts: Research challenges and recent progress, *Interface*, 32(6), 58–76.

Chan, F.T.S., Lau, K.W., Chan, P.L.Y., and Choy, K.L. (2006). Two-stage approach for machine-part grouping and cell layout problems, *Robotics and Computer-Integrated Manufacturing*, 22, 217–238.

Chien, T.-K. (2004). An empirical study of facility layout using a modified SLP procedure, *Journal of Manufacturing Technology Management*, 15(6), 455–465.

Drira, A., Pierreval, H., and Gabou, S.H. (2007). Facility layout problems: A survey, *Annual Reviews in Control*, 31(2), 255–267.

Foulds, L.R. and Robinson, D.F. (1978). Graph theoretic heuristics for the plant layout problem, *International Journal of Production Research*, 16, 27–37.

Garey, M.R. and Johnson, D.S. (1979). *Computers and Intractability: A Guide to the Theory of NP-Completeness*. W.H. Freeman, New York.

Giffin, J.W., Watson, K., and Foulds, L.R. (1995). Orthogonal layouts using the deltahedron heuristic, *Australasian Journal of Combinatorics*, 12, 127–144.

Han, K.H., Bae, S.M., Choi, S.H., Lee, G., and Jeong, D.M. (2011). Parametric layout design and simulation of flexible manufacturing system. *Proceedings of the 10th WSEAS International Conference on System Science and Simulation in Engineering*, Penang, Malaysia, October 3–5, 2011, pp. 94–99.

Han, K.H., Bae, S.M., and Jeong, D.M. (2012). A decision support system for facility layout changes. In: *Latest Trends in Information Technology*, Anderson, D., Yang, H.-J., and Varacha, P. (eds.), Wseas Press, Vienna, Austria, pp. 79–84.

Heragu, S.S. (1997). *Facilities Design*. PWS Publishing, Boston, MA.

Heragu, S.S. and Kusiak, A. (1991). Efficient models for the layout design problem, *European Journal of Operational Research*, 53, 1–13.

Koopmans, T.C. and Beckmann, M. (1957). Assignment problems and the location of economic activities, *Econometrica*, 25(1), 53–76.

Lee, Y.H. and Lee, M.H. (2002). A shape-based block layout approach to facility layout problems using hybrid genetic algorithm, *Computers & Industrial Engineering*, 42, 237–248.

Mu-jing, Y. and Gen-gui, Z. (2005). Method of systematic layout planning improved by genetic algorithm and its application to plant layout design, *Journal of East China University of Science and Technology*, 31(3), 371–375.

Muther, R. (1973). *Systematic Layout Planning*, 2nd edn. Cahners Books, Boston, MA.

Nadler, G. (1961). *Work Design: A Systems Concept*. Richard D. Irwin, Inc., Homewood, IL.

Nagapak, N. and Phruksaphanrat, B. (2011). Cellular manufacturing layout design and selection: A case study of electronic manufacturing service plant. *Proceedings of the International Multiconference of Engineers and Computer Scientists*, Kowloon, Hong Kong, Vol. 2, March 16–18, 2011.

Peters, B.A. and Yang, T. (1997). Integrated facility layout and material handling system design in semiconductor fabrication facilities, *IEEE Transactions on Semiconductor Manufacturing*, 10(3), 360–369.

Shayan, E. and Chittilappilly, A. (2004). Genetic algorithm for facilities layout problems based on slicing tree structure, *International Journal of Production Research*, 42(19), 4055–4067.

Singh, S. P. and Sharma, R.R.K. (2006). A review of different approaches to the facility layout problems, *International Journal of Advanced Manufacturing Technology*, 30(5/6), 425–433.

Tompkins, J.A., White, J.A., Bozer, Y.A., Frazelle, E.H., Tanchoco, J.M., and Trevino, J. (1996). *Facilities Planning*. Wiley, New York.

Van Donk, D.P. and Gaalman, G. (2004). Food safety and hygiene: Systematic layout planning of food processes, *Chemical Engineering Research and Design*, 82(A11), 1485–1493.

Weiss, H.J. and Gershon, M.E. (1993). *Production and Operations Management*, 3rd edn. Allyn & Bacon, Boston, MA.

Yang, T. and Hung, C.-C. (2007). Multiple-attribute decision making methods for plant layout design problem, *Robotics and Computer-Integrated Manufacturing*, 23, 126–137.

Yang, T., Peters, B.A., and Tu, M. (2005). Layout design for flexible manufacturing systems considering single-loop directional flow patterns, *European Journal of Operational Research*, 164, 440–455.

Yang, T., Su, C.-T., and Hsu, Y.-R. (2000). Systematic layout planning: A study on semiconductor wafer fabrication facilities, *International Journal of Operation Production Management*, 20(11), 1359–1371.

11

LOCATION PROBLEMS WITH DEMAND REGIONS

DERYA DİNLER, MUSTAFA KEMAL TURAL, AND CEM İYİGÜN

Contents

11.1 Introduction

Facility location problems involve strategic decisions requiring large investments and long-term plannings. They have been extensively studied by researchers from a variety of disciplines, like geographers and marketing and supply chain specialists.

The facility location problem is to locate q serving facilities for m demanding entities and to allocate the demanding entities to the facilities so as to optimize a certain objective such as minimizing transportation cost or the distance to the farthest entity. Most facility location problems are combinatorial in nature and challenging to solve to optimality. Locations of warehouses, hospitals, retail outlets, radar beams, and exploratory oil wells are some of the application areas of the facility location problems. These problems can differ in several ways including the objective aimed, the number of facilities to locate, and the solution space in which the problem is defined. The facility location problem is called as a discrete facility location problem if there are a finite number of candidate facility locations. If the facilities can be placed anywhere in some continuous region, then the problem is called as a continuous facility location problem. When q is equal to 1, the problem is called as a single-facility location problem; otherwise, it is called multiple-facility location problem. More details about facility location problems particularly about their classification can be found in [17,19,20].

In location theory, customers are generally assumed as fixed points in space. When the sizes of the customers are relatively small with respect to the distances between facilities and customers, this assumption can be justified. Otherwise, it would be better to treat customers as groups of points or regions with density functions or distributions representing the demand over the regions.

Here, we restrict ourselves to the case where each demanding entity is represented by a region. It would be more appropriate to represent a demanding entity as a region instead of a fixed point, when

1. The size of the demanding entity is not negligible with respect to the distances in the problem
2. The location of the demanding entity follows a bivariate distribution on the plane
3. The number of demanding entities is so large that they may be clustered into regions instead of treating each one separately

The concept of demand spreading over an area appears in several applications. For example, first consider the problem of locating a fire station that will serve forests. If each forest is represented as a point by its center and a fire bursts out at an area far from the center, it may take more than the estimated time for the firefighters to reach the fire area. In such cases, representing forests as demand areas would be more meaningful as in case (1). Second, consider establishing mobile headquarters or mobile health centers. The location of each unit will follow a bivariate distribution. The decision of where to place a facility to serve these units should consider each as a region along with a density function that represents the likelihood of the presence of the unit as in case (2). Third, consider the problem of waste collection from many districts. Waste collection center should be located by treating each district as groups of demand points (private residences) or regions as in case (3).

In this chapter, we consider continuous facility location problems where the demanding entities are represented as regions in the plane. The chapter is organized as follows. In Section 11.2, we introduce two solution approaches for continuous facility location problems with demand points that are commonly used in solving the problems with demand regions. In Section 11.3, we model 12 continuous location problems with demand regions. We focus on one of the problems and discuss several solution approaches. We provide a brief literature review on some of the remaining problems.

11.2 Location Problems with Demand Points

The Weber problem is a well-known continuous facility location problem [25]. It is to find a center (x^*, y^*) so as to minimize the sum of weighted Euclidean distances between this center and m fixed points (demand points) with given coordinates (a_i, b_i), $i = 1, 2, \ldots, m$. Each point i is associated with a positive weight w_i. The problem can be formulated as follows:

$$\text{Minimize}_{x,y} \quad W(x, y) = \sum_{i=1}^{m} w_i d_i(x, y), \qquad (11.1)$$

where

$$d_i(x, y) = \sqrt{(x - a_i)^2 + (y - b_i)^2}. \qquad (11.2)$$

A common approach to solve the Weber problem is the Weiszfeld procedure [28]. It is an iterative method that expresses and updates the facility location as a convex combination of the locations of the customers.

Another important continuous facility location problem is the location allocation (LA) problem, which is a generalization of the Weber problem to the case of multiple facilities. Given the location of a set of m demand points, LA problem is to find the locations of q facilities and to allocate the demand points to the facilities while minimizing the total distance between the demand points and the facilities they are allocated to. The alternate location allocation (ALA) heuristic developed by Cooper [12] for the LA problem is one of the most commonly used schemes in the multifacility location literature. ALA heuristic mainly depends on two simple problems: (1) given the facility locations, determine the allocations of the demand points, and (2) given the allocations of the points, find the locations of the facilities. These problems can be solved in an iterative manner that starts with a set of initial facility locations and generates q subsets of demand points by allocating each point to one of the facilities. Then, for each subset, a single-facility location problem is solved, and each demand point is allocated to the nearest facility hence generating new subsets of demand points. Iterations are repeated until a stopping criterion is achieved.

11.3 Location Problems with Demand Regions

When demand regions are in consideration, an important aspect is the way of measuring the distance between a demand region and a facility. Usually, there are three ways to define the distance [22]:

1. Maximum distance
2. Minimum distance
3. Average distance

Using one of these distance measures, a facility location problem can be modeled with several objectives. Here, we consider four different objectives: (1) minimize the sum of distances, (2) minimize the maximum distance, (3) maximize the sum of distances, and (4) maximize the minimum distance. In total, there are 12 possible problems. Assuming each demand region has a finite number of demand points, all these problems are formulated using the following notation:

- $q \geq 1$: number of facilities
- $m \geq 1$: number of demand regions

- $x_i \in \mathfrak{R}^2$: location of the ith facility, $i = 1, \ldots, q$
- $k_j \geq 1$: number of demand points in the jth demand region, $j = 1, \ldots, m$
- $K_j := \{1, \ldots, k_j\}, j = 1, \ldots, m$
- $s_j^k \in \mathfrak{R}^2$: location of the kth demand point of the jth demand region, $j = 1, \ldots, m, k \in K_j$
- $w_j > 0$: weight of the jth demand region, $j = 1, \ldots, m$
- d: a distance metric on $\mathfrak{R}^2$ that measures the distance between any two points is $\mathfrak{R}^2$

The maximum, minimum, and average distances from the jth demand region to the *closest facility* denoted, respectively, by d_{max}^j, d_{min}^j, and d_{avg}^j are calculated as

$$d_{max}^j = \min_{i=1,\ldots,q} \left\{ \max_{k \in K_j} d\left(s_j^k, x_i\right) \right\}, \tag{11.3}$$

$$d_{min}^j = \min_{i=1,\ldots,q} \left\{ \min_{k \in K_j} d\left(s_j^k, x_i\right) \right\}, \tag{11.4}$$

$$d_{avg}^j = \min_{i=1,\ldots,q} \left\{ \frac{1}{k_j} \sum_{k=1}^{k_j} d\left(s_j^k, x_i\right) \right\}. \tag{11.5}$$

All of the 12 possible problems are summarized in Table 11.1.

The single-facility versions of problems 2, 8, and 9 are equivalent to known problems with demand points [18]. The same does not hold for the multiple-facility versions of these problems. The single-facility version of problem 11 is equivalent to a single-facility location problem where the objective is to maximize a weighted sum of the distances between the facility and all demand points.

Selected solution approaches from the literature for some problems in Table 11.1 will be reviewed in the following sections. Our aim is not to completely review the related literature. We will explain some problems in more detail while giving less or no attention to some others.

Table 11.1   Classification of Problems

NO.	OBJECTIVE	OBJECTIVE FUNCTION
Problems with maximum distance		
1	Minimize the sum of maximum distances	$\text{Minimize}\left\{\sum_{j=1}^{m}w_j d_{max}^j\right\}$
2	Minimize the maximum of maximum distances	$\text{Minimize}\{\max_{j=1,\dots,m}\{w_j d_{max}^j\}\}$
3[a]	Maximize the sum of maximum distances	$\text{Maximize}\left\{\sum_{j=1}^{m}w_j d_{max}^j\right\}$
4[a]	Maximize the minimum of maximum distances	$\text{Maximize}\{\min_{j=1,\dots,m}\{w_j d_{max}^j\}\}$
Problems with minimum distance		
5	Minimize the sum of minimum distances	$\text{Minimize}\left\{\sum_{j=1}^{m}w_j d_{min}^j\right\}$
6	Minimize the maximum of minimum distances	$\text{Minimize}\{\max_{j=1,\dots,m}\{w_j d_{min}^j\}\}$
7[a]	Maximize the sum of minimum distances	$\text{Maximize}\left\{\sum_{j=1}^{m}w_j d_{min}^j\right\}$
8[a]	Maximize the minimum of minimum distances	$\text{Maximize}\{\min_{j=1,\dots,m}\{w_j d_{min}^j\}\}$
Problems with average distance		
9	Minimize the sum of average distances	$\text{Minimize}\left\{\sum_{j=1}^{m}w_j d_{avg}^j\right\}$
10	Minimize the maximum of average distances	$\text{Minimize}\{\max_{j=1,\dots,m}\{w_j d_{avg}^j\}\}$
11[a]	Maximize the sum of average distances	$\text{Maximize}\left\{\sum_{j=1}^{m}w_j d_{avg}^j\right\}$
12[a]	Maximize the minimum of average distances	$\text{Maximize}\{\min_{j=1,\dots,m}\{w_j d_{avg}^j\}\}$

[a] The solution space of these problems should be restricted, as otherwise they would not have a finite optimal objective function value.

11.3.1 *Problems with Maximum Distance*

For the problems where the worst-case scenarios are important such as the location of emergency facilities, for example, fire stations, police stations, and hospitals, the maximum distance is commonly used. In such problems, the demand regions can be taken as closed convex polygons as the farthest point will occur at a corner of the convex hull of the demand points in a region (in cases with finitely many demand points). The problems having regions with an infinite number of demand points, for example, ellipsoids, can be handled by approximating the regions with polygons. In this subsection, the regions are all assumed to be closed convex polygons.

11.3.1.1 *Single-Facility Minisum Problem with Euclidean Maximum Distance* The single-facility version of this problem with the Euclidean norm can be modeled as a second-order cone programming (SOCP) problem [16]. It can be solved in the worst case in time $O(m^2N^{3/2})$, where m is the number of demand regions and N is the total number of corners in all the regions. Note that the problems including closed circular regions in addition to polygonal ones can also be directly handled (without a polygonal approximation) by an SOCP formulation. SOCP problems can be solved in polynomial time [1,23], and several efficient software packages have been developed to solve such problems [14].

Jiang and Yuan [21] studied the same problem considering closed convex polygonal and circular demand regions. The difficulty of solving this problem is the discontinuity of the farthest points. They, therefore, partitioned the plane into polygonal fixed regions in such a way that within the interior of each fixed region, there is a single farthest point of each demand region. So, the problem is turned into a set of Weber problems with the additional constraint that the facility has to be confined within a certain polygon. Solving such constrained Weber problems, they find the optimal solution. Weaknesses of this approach are twofold: constructing the fixed regions and possibility of solving a large number of constrained Weber problems. Even in the case with r rectangular regions, the number of fixed regions can be as large as $2r^2 + r + 1$, see [21]. The authors proposed an approach to discard some of the fixed regions without solving the associated constrained Weber problem. Since the computational time of this algorithm is highly dependent on the starting fixed region, the authors also proposed a heuristic for the initialization.

Drezner and Wesolowsky [18] proposed another approach for the same problem. They did not explicitly construct the fixed regions. Instead, they dealt with the discontinuity of the farthest points by an algorithm whose iterations use the Weiszfeld procedure repeatedly together with golden section search as needed.

All three methods explained so far solve the problem to optimality. SOCP formulation is very easy to implement and can be used to solve small- or medium-size instances. As there has not been any computational comparison of the aforementioned methods, it is not known which one would perform the best for large-size instances.

11.3.1.2 Other Minisum Problems with Maximum Distance In [18], authors also studied a single-facility minisum problem with maximum distance using rectilinear norm. They showed that the problem can be formulated as a linear programming problem.

In [22], Jiang and Yuan considered a multiple-facility version of the problem in [21], that is, problem 1 with the Euclidean norm. They developed an ALA heuristic. In the location step, they solved constrained Weber problems as in [21]. Since the number of constrained Weber problems to solve may be too large, the authors proposed a version of Barzilai–Borwein gradient method and proved its convergence. In the allocation step, each region is assigned to the closest facility. They demonstrated the efficiency of their method with a numerical study.

In [15], authors considered a multiple-facility version of the problem using the squared Euclidean norm. They modeled the problem as a mixed-integer SOCP problem. Since this formulation is weak, they proposed three heuristics applicable to general polygons. Two of them are ALA heuristics. Differently, the third one uses a smoothing strategy to turn the problem into an unconstrained nonlinear problem that is then solved with a quasi-Newton algorithm. The authors also proposed a special heuristic for the case where the demand regions are of rectangular shape with sides parallel to the standard coordinate axes.

11.3.2 Problems with Minimum Distance

For the problems where the flow from/to the facilities will enter/leave the given demand area at the closest point (i.e., drop-off and take-off points), the minimum distance is commonly used. The internal distribution costs within the demand area are usually not considered in such problems.

11.3.2.1 Minisum Problems with Minimum Distance   The single-facility version of this problem with closed convex demand regions (not necessarily polygonal regions) is studied in [7]. They proposed an iterative algorithm starting with an arbitrary initial facility. In each iteration, the closest point for each demand region is found, and those points are treated as fixed points replacing the associated demand regions. The resulting Weber problem is solved with the Weiszfeld algorithm. Convergence properties of the algorithm and modifications in some special cases are discussed in detail in [7].

In [5], authors considered both the regional demands and regional facilities. Their objective was to locate a facility in order to minimize the sum of the distances from the closest point in the facility to the closest point in the demand areas. They proved that when the demand regions and the facility are closed convex regions, the distance is a convex function of a defined center of the facility for any norm. Therefore, objective function of the problem is convex, and a classical descent method to find global optimum can be used. However, authors stated that the calculations of step size and descent directions are not easy because of the discontinuity in derivatives and unobtainability of the closest distance in an explicit form. They overcame these difficulties in some special cases by using the rectilinear norm. They presented a solution approach for the case where both the demand and the facility regions are of rectangular shape. The results obtained for the single-facility case with rectangular regions and facility were extended to the multiple-facility case.

In [26], authors represented the demanding entities as convex sets of points. The objective is to minimize an increasing convex function of the minimum distances between the facility and the demand regions. The single-facility minisum problem with minimum Euclidean distances is a special case of the considered problem. The geometrical characterization of the set of optimal solutions is presented by using tools from convex analysis. Moreover, a constructive approach is developed for the case with polyhedral norms.

11.3.2.2 Minimax Problems with Minimum Distance The single-facility minimax problem with Euclidean minimum distance is also a special case of the problem considered in [26]. Same problem with some distance measures was also studied in [6]. They developed a procedure based on the iso-contours. It was shown that the proposed methodology can lead to efficient solution methods in some special cases, for example, rectangular regions with rectilinear distance.

11.3.3 Problems with Average Distance

For the problems where the distances from the facility to each demand point in the region are important, the average distance is commonly considered. It is also generally used in problems where the demand points are represented as random vectors. Average distance has been more extensively used in the literature than the other distance definitions.

Problems with average distances generally require evaluation of complex integral expressions. In [27], Stone gave the explicit expressions and approximate power series for four common cases, namely, rectangular and circular demand regions with both rectilinear norm and Euclidean norm. For each case, expressions for the average distance between the demand region and the facility at the center, interior or exterior of the demand regions, are presented.

11.3.3.1 Minisum Problems with Average Distance Love [24] considered the situation in which the number of demand points is too large to treat each of them separately. He grouped the demand points into rectangular regions. His objective was to find the location of a facility so as to minimize total expected Euclidean distances between the rectangular regions and the facility. He developed a response surface technique utilizing a gradient reducing process.

In [3], the authors extended Love's study [24]. They replaced the demand regions, which are not necessarily rectangular and uniformly distributed, with the centroids of the regions. Their method can be used for demand regions having a geometric shape with an easily found centroid. Comparing with the study in [24], they computationally showed that their method is faster for the rectangular regions with uniformly distributed demand.

In [13], the location of each demanding entity is assumed to be a random variable having a bivariate normal distribution with zero correlation. The author aimed to locate a single facility to minimize the sum of the expected Euclidean distances between the demanding entities and the facility. It was proven that the objective function of the problem is strictly convex and an iterative algorithm for the problem was proposed.

In [2], the problem of locating one or more facilities to serve existing rectangular regions was studied where the rectilinear norm was used. The authors proposed a gradient-free direct search method for the problem and a heuristic for the initialization. Their method converged experimentally, but no formal proof of the convergence was given.

Carrizosa et al. [8] discussed the similarities and differences between a generalized Weber problem with demanding entities represented by density functions and its point version. They showed that when the probability distributions of the demanding entities are absolutely continuous, gradient descent algorithms can be used instead of evaluating complex expectations. The authors also proved that the problem has a unique optimal solution in some special cases.

In [9], the authors approximated demand regions with simpler regions while keeping an approximation error under control. For example, an elliptical region can be approximated by an n-sided polygon where the approximation error gets smaller with larger values of n. For polygonal approximations, the triangles constructed with the corners of the polygons are used to calculate the expected distances. Using this idea, they proposed an algorithm whose running time increases with the number of sides of the approximation polygons. However, they obtained promising results even when the number of sides of the approximation polygon is not too large.

Chen [11] proposed a Weiszfeld-like iterative approach to locate a single facility that serves circular demand regions using the Euclidean norm. He recommended taking a weighted average of the centers of the circular demand regions as a starting point for the approach.

Cavalier et al. [10] studied the problem of minimizing the sum of weighted expected distances between convex polygonal demand regions having uniform demand and a single facility (or multiple ones). For the single-facility case, an iterative algorithm that uses the Weiszfeld technique was developed and the convergence of the algorithm was proven. For the multiple-facility case, an ALA heuristic was presented. This algorithm was also convergent, but the global optimality was not guaranteed.

11.3.3.2 Minimax Problems with Average Distance   In [18], authors also studied a single-facility minimax problem with average distance using the Euclidean norm. An Elzinga–Hearn-type algorithm was proposed.

11.3.4 Problems on Networks

All the papers mentioned until now studied location problems where both the demanding entities and the facilities lie on the plane. In [4], the authors studied location problems on a network. They investigated nine different problems (as in Table 11.1 except problems 3, 7, and 11) where the demand points are clustered into groups. For each problem, a set of potential locations of the facility(s) are derived. For the multiple-facility case, heuristics (tabu search and simulated annealing) were proposed.

11.4 Conclusions

Classical location problems with demand points have been studied for a long time. Recently, new variants of facility location problems with demand regions have attracted the researchers. In this chapter, several facility location problems with demand regions are reviewed and selected solution approaches from the literature are discussed. The papers reviewed are summarized in terms of the type of the problems studied, number of facilities located, type of the demanding entities, distance measure used, and the solution approaches, and presented in Table 11.2.

Table 11.2 Summary of Papers Reviewed

REFERENCES	PROBLEM TYPE (SEE TABLE 11.1)	NUMBER OF FACILITIES	DEMANDING ENTITY	DISTANCE MEASURED BY	SOLUTION APPROACH
[2]	9	Both single and multiple	Rectangular region	Rectilinear norm	Gradient-free direct search method
[3]	9	Single	A geometric shape with an easily found centroid	Euclidean norm	Response surface method
[4]	1, 2, 4, 5, 6, 8, 9, 10, 12	Both single and multiple	Cluster of points on networks	—	Heuristics (improved greedy algorithm, tabu search, and simulated annealing)
[5]	5	Single	General convex region	Arbitrary norm	General convexity results are derived
[5]	5	Both single and multiple	Rectangular region	Rectilinear norm	Reduced the problem to standard minisum problem
[6]	6	Single	Rectangular region	Rectilinear norm	Descent procedure based on iso-contours
[7]	5	Single	Closed convex region	Euclidean norm	Iterative algorithm using Weiszfeld technique
[8]	9	Single	Random variable with a bivariate distribution	A gauge	Properties of the problem are discussed
[9]	9	Single	Bounded set of points	A gauge	Properties of the problem are discussed
[10]	9	Single	Convex polygonal region	Euclidean norm	Iterative algorithm using Weiszfeld technique
[10]	9	Multiple	Convex polygonal region	Euclidean norm	Alternate location allocation heuristic
[11]	9	Single	Circular region	Euclidean norm	Weiszfeld-like iterative approach
[13]	10	Single	Bivariate normally distributed random variable	Euclidean norm	An iterative algorithm

(Continued)

Table 11.2 (*Continued*) Summary of Papers Reviewed

REFERENCES	PROBLEM TYPE (SEE TABLE 11.1)	NUMBER OF FACILITIES	DEMANDING ENTITY	DISTANCE MEASURED BY	SOLUTION APPROACH
[15,16]	1	Multiple	Closed convex polygonal region	Euclidean norm	Heuristics (two alternate location allocation heuristics, an iterative algorithm based on a smoothing strategy)
[15,16]	1	Multiple	Rectangular region with sides parallel to coordinate axes	Euclidean norm	Alternate location allocation heuristic
[15,16]	1	Single	Closed convex polygonal region	Euclidean norm	Second-order cone programming formulation
[18]	1	Single	Closed convex polygonal or circular region	Euclidean norm	Iterative algorithm using Weiszfeld technique
[18]	1	Single	Closed convex polygonal or circular region	Rectilinear norm	Linear programming formulation
[18]	10	Single	Closed convex polygonal or circular region	Euclidean norm	Elzinga–Hearn-type algorithm
[21]	1	Single	Closed convex polygonal or circular region	Euclidean norm	Divided the solution space into fixed regions, and solved several constrained Weber problems
[22]	1	Multiple	Closed convex polygonal or circular region	Euclidean norm	Alternate location allocation heuristic
[24]	9	Single	Rectangular region	Euclidean norm	Response surface method
[26]	5	Single	Convex sets of points	Polyhedral norm	A constructive approach

References

1. Alizadeh, F., Goldfarb, D. 2003. Second-order cone programming, *Mathematical Programming* 95:3–51.
2. Aly, A.A., Marucheck, A.S. 1982. Generalized Weber problem with rectangular regions, *The Journal of the Operational Research Society* 33:983–989.
3. Bennett, C.D., Mirakhor, A. 1974. Optimal facility location with respect to several regions, *Journal of Regional Science* 14:131–136.
4. Berman, O., Drezner, Z., Wesolowsky, G.O. 2001. Location of facilities on a network with groups of demand points, *IIE Transactions* 33:637–648.
5. Brimberg, J., Wesolowsky, G.O. 2000. Note: Facility location with closest rectangular distances, *Naval Research Logistics* 47:77–84.
6. Brimberg, J., Wesolowsky, G.O. 2002. Locating facilities by minimax relative to closest points of demand areas, *Computers and Operations Research* 29:625–636.
7. Brimberg, J., Wesolowsky, G.O. 2002. Minisum location with closest Euclidean distances, *Annals of Operations Research* 11:151–165.
8. Carrizosa, E., Conde, E., Munoz-Marquez, M., Puerto, J. 1995. The generalized Weber problem with expected distances, *RAIRO Operations Research* 29:35–57.
9. Carrizosa, E., Munoz-Marquez, M., Puerto, J. 1998. The Weber problem with regional demand, *European Journal of Operational Research* 104:358–365.
10. Cavalier, T.M., Sherali, H.D. 1986. Euclidean distance location-allocation problems with uniform demands over convex polygons, *Transportation Science* 20:107–116.
11. Chen, R. 2001. Optimal location of a single facility with circular demand areas, *Computers and Mathematics with Applications* 41:1049–1061.
12. Cooper, L. 1964. Heuristic methods for location-allocation problems, *SIAM Review* 6:37–53.
13. Cooper, L. 1974. A random locational equilibrium problem, *Journal of Regional Sciences* 14:47–54.
14. CVX Research, Inc. April 2011. CVX: Matlab software for disciplined convex programming, version 2.0. http://www.cvxr.com/cvx (accessed August 7, 2014).
15. Dinler, D., Tural, M.K., İyigün, C. 2013. Heuristics for a continuous multi-facility location problem with demand regions, Submitted for publication.
16. Dinler, D. 2013. Heuristics for a continuous multi-facility location problem with demand regions, Master's thesis. Middle East Technical University, Ankara, Turkey.
17. Drezner, Z. (ed.). 1995. *Facility Location: A Survey of Applications and Methods*. Springer, New York.
18. Drezner, Z., Wesolowsky, G.O. 2000. Location models with groups of demand points, *INFOR* 38:359–372.
19. Drezner, Z., Hamacher, H.W. (eds.). 2002. *Facility Location: Applications and Theory*. Springer, Berlin, Germany.

20. Farahani, R.Z., Hekmarfar, M. (eds.). 2009. *Facility Location: Concepts, Models, Algorithms and Case Studies.* Springer, Berlin, Germany.
21. Jiang, J., Yuan, X. 2006. Minisum location problem with farthest Euclidean distances, *Mathematical Methods of Operations Research* 64:285–308.
22. Jiang, J., Yuan, X. 2012. A Barzilai-Borwein-based heuristic algorithm for locating multiple facilities with regional demand, *Computational Optimization and Applications* 51:1275–1295.
23. Lobo, M.S., Vandenberghe, L., Boyd, S., Lebret, H. 1998. Applications of second-order cone programming, *Linear Algebra and Its Applications* 284:193–228.
24. Love, R.F. 1972. A computational procedure for optimally locating a facility with respect to several rectangular regions, *Journal of Regional Sciences* 12:233–242.
25. Luangkesorn, L. Weber problem [PDF document]. Retrieved from: http://www.pitt.edu/~lol11/ie1079/notes/ie2079-weber-slides.pdf (accessed September 24, 2013).
26. Nickel, S., Puerto, J., Rodriguez-chia, A. 2003. An approach to location models involving sets as existing facilities, *Mathematics of Operations Research* 28:693–715.
27. Stone, R.E. 1991. Some average distance results, *Transportation Science* 25:83–91.
28. Weiszfeld, E. 1937. Sur le point par lequel la somme des distances de n points donnés est minimum, *Tohoku Mathematical Journal* 43:355–386.

12

A NEW APPROACH FOR SYNCHRONIZING PRODUCTION AND DISTRIBUTION SCHEDULING

Case Study

E. GHORBANI-TOTKALEH, M. AMIN NAYERI, AND M. SHEIKH SAJADIEH

Contents

12.1 Introduction

Production and distribution are two important processes in supply chain. In the recent two decades, many works have been done on integrated production–distribution models in strategic and tactical planning fields (Bilgen and Ozkarahan, 2004, Chen, 2004, Goetschalcks et al., 2002). These articles consider inventory decisions to link these two parts of supply chain in decision problems. Contrary to this, integrated production and distribution scheduling problem has come into consideration, recently. For having a global optimal solution for our scheduling problem in production and distribution parts of supply chain, we must consider these two parts integrated. It is very important in make-to-order products because for this kind of products like time-sensitive products, we should find an optimal solution to have on-time delivery with the minimum cost. Formerly, scheduling of production and distribution parts was done consecutively, in other words, the output of production scheduling decision was an input of distribution scheduling part. It has been shown that this sequential approach has suboptimal solution (Chen and Vairaktarakis, 2005, Pundoor and Chen, 2005). Thus, integrated production–distribution scheduling is optimal and more beneficial.

Moreover, the integrated approach for scheduling can reduce total cost and increase the customer service level due to lead time reduction. We can also estimate lead time or due date more accurately if considering delivery and production schedules integrated.

Production and distribution scheduling problems can be seen separately in many articles in the past decades (Ball et al., 1995, Pinedo, 2002). By considering production and distribution integrative for scheduling, we have a relatively new area that has been studied in the past decade. This stream of literature can be called integrated production and outbound distribution scheduling (IPODS) problems (Chen, 2010).

The rest of the chapter is organized as follows. In Section 12.2, we review related literature and existing models in IPODS area. Notation and description of our specific problem in real world are given in Section 12.3. In Section 12.4, after modeling our problem mathematically, the detailed solution method is presented. Numerical results for evaluating our proposed solution method and for gaining insights are presented in Section 12.5. Conclusions are provided in Section 12.6.

12.2 Literature Review

Zhi-Long Chen (2010) proposed a five-field notation, $\alpha|\beta|\pi|\delta|\gamma$, to represent most IPODS models. In this notation, α describes the machine configuration in the production part, β specifies restrictions about orders, and γ describes the objective function. These three factors are used in production scheduling models (Lawler et al., 1993,

Table 12.1   Types of Machine Configuration (α)

VALUE	DESCRIPTION
1	Single machine
Pm	Parallel machine
Fm	Flow shop
Bm	Bundling
$F(m_1, m_2)$	Two-stage flexible flow shop
MP	Multiple plants

Table 12.2   Restrictions and Constraints on Order Parameters (β)

VALUE	DESCRIPTION
r_j	Orders have unequal release dates
$d_j \equiv d$	Orders have a common due date d
$\overline{d_j}$	Each order j has a deadline
$[a_j, b_j]$	Order j must be delivered to its customer within this time window
fd_j	Each order j has a fixed delivery time
s_{ij}	There are sequence-dependent setup times between different orders
Prec	Orders have precedence constraints between them
Pmtn	Order processing can be preempted and resumed later
Pickup	Orders must be picked up from the customer before they can be processed
No wait	Each order should be processed without idle time from one machine to the next
r-a	One machine has a known unavailable period of time
$\sum D_j \leq D_0$	Total delivery time of the orders cannot exceed a specified deadline

Pinedo, 2002). Values that can be taken for α and β are presented in Tables 12.1 and 12.2, respectively. γ shows the objective functions like customer service level, total cost, and total revenue.

There are two other factors in the notation proposed by Chen (2010), π and δ, which specify the delivery process and the number of customers, respectively. The first factor, π, shows the features of vehicles by $V(x,y)$ (number and capacity of vehicles) and delivery method, that is, there are two parameters in this notation. The value of each parameter can be seen in Table 12.3. Most of existing models consider homogeneous vehicles, and this notation shows these models only.

The second factor, δ, defines the number of customers, and it is 1 (one customer), k (k customers who are fixed in each scheduling horizons), or n (different number of customers in different scheduling horizons).

Table 12.3   Delivery Characteristics (π)

PARAMETER	VALUE	DESCRIPTION	WHAT IT SHOWS
X	1	Single-delivery vehicle	Number of vehicles
	υ	Number of available vehicles is finite or limited	
	∞	There are sufficient number of vehicles or infinite	
Y	1	Each shipment can deliver only one unit of product	Capacity of vehicles
	c	Each shipment can deliver up to c orders (orders have an equal size)	
	∞	Each shipment can deliver infinite number of order	
	Q	Each shipment can deliver at most Q units (orders have different sizes)	
	iid	Individual and immediate delivery	Methods of distribution
	direct	Batch delivery by direct shipping	
	routing	Batch delivery with routing	
	fdep	Shipping with fixed delivery departure date	
	split	An order can be split to be delivered in multiple shipments	

Table 12.4 reviews the recent articles considering IPODS models.

12.3 Problem Description

In this part, we want to model a problem in a company. We consider each 30 minutes, to be one time scale. Thus, there are 16 time units per day, as the time horizon of scheduling. In our case study, orders need to be processed on a single machine ($\alpha = 1$). Each customer has a time window such as $[a_i, b_i]$ for the delivery of their demands. If the delivery is received earlier (delivery time < a_i) or later (delivery time > b_i), a penalty cost for per unit of time delay is considered. So orders have specific time window for delivery ($\beta = [a_i, b_i]$). In this problem, there are three vehicles of two different types with different capacities, so they have different transportation costs. The specific notation assignment for different vehicles is not mentioned by Chen (2010). We can show it as $V_1(1,Q_1)$, $V_2(2,Q_2)$. A routing method for the delivery of our orders is used, as this method can

Table 12.4 IPODS Models in Existing Articles

REFERENCES	α	β	π	δ	γ	SOLUTION METHOD
Chen and Pundoor (2009)	1	$\overline{d}_j$	V(∞,Q), direct, split	1	TC	Heuristic algorithm
Chen and Lee (2008)	1		V(∞,c), direct	k	$\sum w_j D_j + TC$	Polynomial-time algorithm
Zhong et al. (2007)	1		V(1,Q), direct	1	D_{max}	Heuristic algorithm
Li and Ou (2007)	MP		V(∞,c), direct	1	$\sum D_j + TC$	Heuristic algorithm
Averbakh and Xue (2007)	1	r_j, pmtn	V(∞,∞), direct	1,k	$\sum D_j + TC$	Online algorithm
Li and Vairaktarakis (2007)	B2		V(∞,c), direct/routing	k	$\sum D_j + TC$	Heuristic algorithm, Polynomial-time approximation scheme
Ji et al. (2007)	1		V(∞,∞), direct	1	$\sum w_j D_j + TC$	Polynomial-time exact algorithm
He et al. (2006)	1		V(1,Q), direct	1	D_{max}	Heuristic algorithm
Dawande et al. (2006)	1	s_{ij}	V(∞,1), iid	k	$\alpha C_{max} + (1-\alpha)L_{max}$	Exact algorithm
Chen and Pundoor (2006)	MP		V(∞,c), direct	1	$\sum D_j + TC + PC$	Heuristic algorithm
Pundoor and Chen (2005)	1		V(∞,c), direct	k	$L_{max} + TC$	Heuristic algorithm
Li and Ou (2005)	1	pickup	V(1,c), direct	1	D_{max}	Heuristic algorithm
Wang and Lee (2005)	1	$\overline{d}_j$	V(∞,1), iid	N	TC	Exact algorithm
Chen and Vairaktarakis (2005)	Pm		V(∞,c), direct	1	$D_{max} + TC$	Heuristic algorithm
Chen and Vairaktarakis (2005)	Pm	s_{ij}, d_j	V(∞,c), routing	k	$D_{max} + TC$	Heuristic algorithm
Li et al. (2005)	1		V1,c), routing	k	$\sum D_j$	Polynomial solvable

(Continued)

Table 12.4 (Continued) IPODS Models in Existing Articles

REFERENCES	α	β	π	δ	γ	SOLUTION METHOD
Garcia and Lozano (2005)	Pm	$[a_j, b_j]$	$V(\upsilon,1)$, iid	n	$\sum R_j$	Tabu search
Chang and Lee (2004)	1		$V(1,Q)$, direct/routing	1, 2	D_{max}	Heuristic algorithm
Garcia et al. (2004)	MP	fd_j	$V(\upsilon,1)$, iid	n	$\sum R_j - TC$	Heuristic algorithm
Garcia and Lozano (2004)	Pm	fd_j	$V(\infty,1)$, iid	n	$\sum R_j$	Min cost network flow
Mastrolilli (2003)	1,Pm	r_j	$V(\infty,1)$, iid	n	D_{max}	Polynomial-time approximation scheme
Kaminsky (2003)	Fm	r_j	$V(\infty,1)$, iid	n	D_{max}	Heuristic algorithm
Hall and Potts (2003)	1		$V(\infty,\infty)$, direct	k	$\sum U_j + TC$	Ordinary NP-hard
Gharbi and Haouari (2002)	Pm	r_j	$V(\infty,1)$, iid	n	D_{max}	Branch-and-bound-based exact algorithm
Liu and Cheng (2002)	1	r_j, s_j, pmtn	$V(\infty,1)$, iid	n	D_{max}	Polynomial-time approximation scheme
Garcia et al. (2002)	Pm	fd_j	$V(\upsilon,1)$, iid	n	$\sum R_j$	Heuristic algorithm
Lee and Chen (2001)	1		$V(\upsilon,c)$, direct	1	D_{max}	
Wang and Cheng (2000)	Pm		$V(\infty,\infty)$, direct	1	$\sum D_j + TC$	Polynomial-time exact algorithm
Van Buer et al. (1999)	1	s_j, d_j	$V(\infty,Q)$, routing	n	$TC+VC$	Heuristic algorithm
Woeginger (1998)	1	s_j	$V(\infty,1)$, iid	n	D_{max}	Polynomial-time approximation scheme
Cheng et al. (1996)	1		$V(\infty,\infty)$, direct	1	$\sum w_j E_j + TC$	Exact algorithm
Woeginger (1994)	Pm		$V(\infty,1)$, iid	n	D_{max}	Heuristic algorithm
Hall and Shmoys (1992)	1	r_j,prec	$V(\infty,1)$, iid	n	D_{max}	Heuristic algorithm

save transportation costs. In each time horizon, there are a number of different customers ($\delta = n$). We need to detect the time of starting production as well as the time of dispatching each vehicle in order to minimize total cost including the travel cost and tardiness and earliness penalty costs. The model can be written as follows:

$$1\left[a_j, b_j\right] V_1\left(1, Q_1\right), V_2\left(2, Q_2\right), \text{routing} \left| n \right| VC + \sum\left(\alpha_i E_i + \beta_i T_i\right)$$

Our innovation in this chapter is to solve a real problem in a company, and our goal is to minimize company's tardiness and earliness penalty costs and its transfer costs.

For modeling this problem, some assumptions need to be made as follows to relax the problem:

1. Each region is considered as a customer, that is, we calculate the total demand of one region and program to send its demand in the best time to minimize the transfer and penalty costs.
2. We relax the service time in the time window, that is, if the service time takes one unit of time and time window is [a,b], then we consider it as [a−1, b−1].
3. The total demand of each region is not over the capacity of all of the vehicles, and thus, at least one vehicle is enough to transfer goods to destinations.
4. All vehicles are in the depot at time zero. Moreover, they do not return to the depot after finishing their job. So each vehicle is available once in each horizon.

The problem is characterized by the following sets and parameters:

- Sets
  $C = \{1, 2, \ldots, n\}$ set of customers in scheduling horizon (1 day)
  $Z = \{0, 1, 2, \ldots, s\}$ set of region zones
  $V = \{1, 2, \ldots, m\}$ set of vehicles
- Decision variables
  $x_{jkk'}$ binary decision variable, taking a value of 1 if vehicle j is used for transporting from region k to k′ and 0 otherwise
  t_j^s dispatching time of vehicle j from depot
  $t_{k'}^p$ production start time for demands of region k′

- Nondecision variables and parameters

  Cap_j capacity for vehicle j

  $[a_i,b_i]$ delivery time window for ith customer

  $c_{jkk'}$ cost of traveling from k to k' by vehicle j'

  q_k volume of the demands of region k

  α penalty per unit time of tardiness

  β penalty per unit time of earliness

  $t_{kk'}$ travel time between k and k'

  p_k production time for demands of region k

  D_{rk} time arriving at region k in the rth order

  τ_{ik} binary variable, taking a value of 1 if customer i is in the region k and 0 otherwise

  T_i tardiness of delivery to customer i

  E_i earliness of delivery to customer i

The mixed-integer nonlinear programming (MINLP) model is formulated as follows:

$$\min \sum_{j=1}^{m}\sum_{k=0}^{s}\sum_{k'=1}^{s} x_{jkk'}\, c_{jkk'} + \sum_{\forall i}(\alpha T_i + \beta E_i) \tag{12.1}$$

$$\text{s.t.} \sum_{j=1}^{m}\sum_{k'=1}^{s} x_{j0k'} \le m \tag{12.2}$$

$$\sum_{j=1}^{m}\sum_{k'=1}^{s} x_{j0k'} \ge 1 \tag{12.3}$$

$$\sum_{j=1}^{m}\sum_{k'=1}^{s} x_{jkk'} \le 1 \quad k \in \{1,\ldots,s\} \tag{12.4}$$

$$\sum_{j=1}^{m}\sum_{k=0}^{s} x_{jkk'} = 1 \quad k' \in \{1,\ldots,s\} \tag{12.5}$$

$$x_{1kk'} + \sum_{l=1}^{s} x_{2k'l} + \sum_{l=1}^{s} x_{3k'l} \leq 1 \quad \forall k; \forall k' \qquad (12.6)$$

$$x_{2kk'} + \sum_{l=1}^{s} x_{1k'l} + \sum_{l=1}^{s} x_{3k'l} \leq 1 \quad \forall k; \forall k' \qquad (12.7)$$

$$x_{3kk'} + \sum_{l=1}^{s} x_{1k'l} + \sum_{l=1}^{s} x_{2k'l} \leq 1 \quad \forall k; \forall k' \qquad (12.8)$$

$$\sum_{k=0}^{s} \sum_{k'=1}^{s} q_{k'} \, x_{jkk'} \leq Cap_j \quad \forall j \in V \qquad (12.9)$$

$$t_j^s = \sum_{k=0}^{s} \sum_{k'=1}^{s} (p_{k'} + t_{k'}^p) x_{jkk'} \quad \forall j \in V \qquad (12.10)$$

$$D_{1k'} = \sum_{j=1}^{m} (t_j^s \, x_{j0k'} + t_{0k'} x_{j0k'}) \quad \forall k' \in \{1,\ldots,s\} \qquad (12.11)$$

$$D_{rk'} = \sum_{j=1}^{m} \sum_{k=1}^{s} (D_{r-1,k} \, x_{jkk'} + t_{kk'} \, x_{jkk'}) \quad r \geq 2 \,, \forall k' \in \{1,\ldots,s\} \qquad (12.12)$$

$$T_i + b_i \geq \sum_{k'=1}^{s} \sum_{r=1}^{s} D_{rk'} \, \tau_{ik'} \quad \forall i \in C \qquad (12.13)$$

$$a_i + E_i \geq \sum_{k'=1}^{s} \sum_{r=1}^{s} D_{rk'} \, \tau_{ik'} \quad \forall i \in C \qquad (12.14)$$

$$x_{jkk'} \in \{0,1\} \,, \quad k = 0,\ldots,s, \, k',r = 1,\ldots,s \,, \, k \neq k' \qquad (12.15)$$

$j = 1, \ldots, m, \, i = 1, \ldots, n$

The objective function (12.1) minimizes the total cost including the travel cost, tardiness and earliness penalty costs. Constraints (12.2) and (12.3) specify that the number of all vehicles, which are traveled from depot to all regions, should be limited in the range of [1, vehicle number]. Constraint (12.4) shows that we cannot send over one vehicle from each region, while constraint (12.5) ensures that every region is only serviced once. Constraints (12.6) through (12.8) show that we enter into and exit from all regions by the same vehicle, and we cannot change the vehicle in regions. It should be noted that we consider these three constraints because in our case study, we have three vehicles. Constraint (12.9) controls the vehicle capacity. Constraint (12.10) calculates t_j^s for each vehicle. Constraints (12.11) and (12.12) calculate the service time for all regions by considering the order of serving. Constraints (12.13) and (12.14) represent tardiness and earliness ranges.

12.4 Solution Approach

In this section, we propose an exact and a heuristic method to solve the problem. Our case study has a small and medium size, so we can solve our problem exactly. However, as Karp (1972) proved that the traveling salesman problem (TSP) is strongly NP-hard, we propose a heuristic algorithm for solving some large size of such problems.

12.4.1 Exact Solution

For solving and testing this model, we identify 20 experiments of the company from 20 days. This company is active in furniture industry. We consider the warehouse of this company as a main depot for production and distribution. The scheduling horizon is assumed to be a working day, starting from 8:00 AM to 4:00 PM. By dividing this time window into 16 parts, each time unit will be 30 min. We consider the process of assembly, disassembly, and packing of the products as manufacturing process. A team of workers do these works together, and can be considered as a single machine. Three different vehicles do the job of transferring products to the customer's place. Each vehicle has a different capacity and different transporting cost, but they have the same speed. Customers can specify their delivery time window. We try to schedule the manufacturing and delivery process to minimize the sum of transportation cost and the penalty of tardiness and earliness of deliveries.

We coded this mathematical model in Microsoft visual studio 2010 and then used its text output for Lingo 11.0, and from all experiments, we can get best answer by the branch-and-bound solver. It is important that for bigger sizes of problems, this exact method cannot give any answer in a reasonable time, so we have to use a heuristic algorithm.

12.4.2 Proposed Algorithm

A heuristic algorithm is proposed for solving the problem, because this routing problem with multiple customers contains the strongly NP-hard TSP (Karp, 1972). As mentioned before, our case study problem is small and medium, and can be solved by the exact method. The proposed algorithm is then appropriate for solving the other cases with larger sizes.

In this algorithm, we use genetic algorithm (GA) because experiences show that population-based algorithms have better proficiency for solving routing and batching problems like our case. After generation and improvement with GA, this algorithm starts local improvement phase by using simulated annealing (SA). After a specific iteration, the algorithm stops and gives the best answer over all iterations. Figure 12.1 shows the flowchart of the proposed algorithm.

To explain the proposed algorithm, some important steps are given as follows:

1. Define chromosomes as can be seen in Figure 12.2.
2. Use partially mapped crossover as shown in Figure 12.3.

   As you can see in Figure 12.3, this operator randomly specifies two points of each parent. From the first parent, the specific part will be copied to the child exactly, and another part of child will come from the second parent randomly.
3. Use replacement operator for mutation. You can see an example of this operator in Figure 12.4.
4. For determining the initial temperature in SA part of the algorithm, we define a linear connection between T_0 and number of regions (s): $T_0 = \theta \cdot s$; where θ is a constant number.

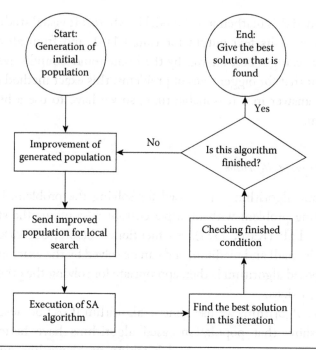

Figure 12.1   Flowchart of solution algorithm.

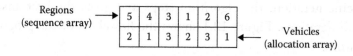

Figure 12.2   A sample of chromosomes in the proposed algorithm.

5. Use geometric approach for the cooling of initial temperature: $T_i = \alpha^i T_0$; where $0.5 \leq \alpha \leq 0.99$.
6. Neighborhood creation is done by using inversion operator for sequence array and insertion operator for allocation array (see Figures 12.5 and 12.6).

For parameter tuning, factorial design is used and three levels of three important parameters in the algorithm are tested. Table 12.5 shows these levels of experiment. By running all $3^3 = 27$ tests, best levels are as determined as follows:

Number of initial population: 60, probability of crossover: 0.6, and ratio of initial temperature: 1.

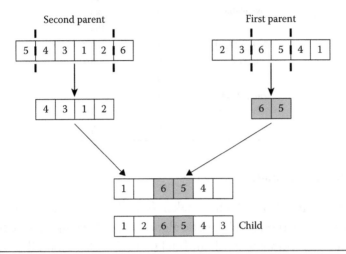

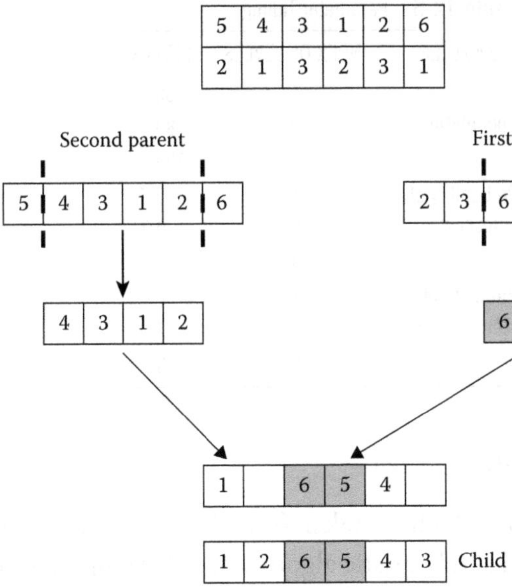

Figure 12.3 Crossover operator in the solution algorithm.

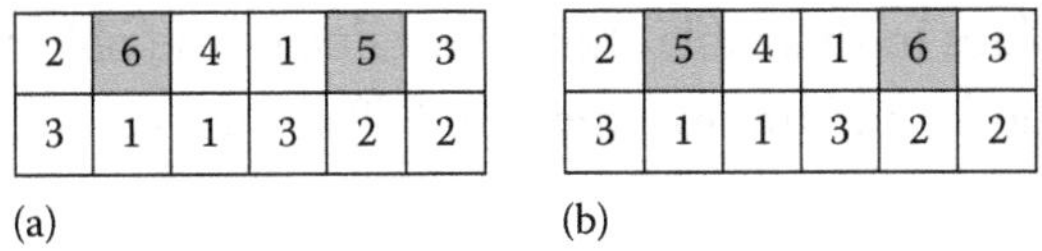

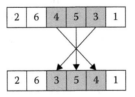

Figure 12.4 Mutation operator in the proposed algorithm: (a) before mutation and (b) after mutation.

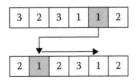

Figure 12.5 Neighborhood creation by inversion operator for sequence array.

Figure 12.6 Neighborhood creation by insertion operator for allocation array.

Table 12.5 Parameter Tuning

PARAMETER	INDEX OF LEVELS	LEVELS
Initial	1	60
population	2	80
	3	100
Probability of	1	0.6
crossover	2	0.7
	3	0.8
Ratio of initial	1	1
temperature	2	2
	3	3

12.5 Numerical Study

Twenty test problems which are taken from 20 working days are solved by the exact method and heuristic algorithm. All examples were tested on a personal computer with an Intel Core i5 CPU, 2.66 GHz, 4 GB RAM. Solutions of exact method and the proposed heuristic algorithm are compared with real practice. Table 12.6 shows results for 20 test problems. Values in this table show the value of the objective function in three methods, and improvement in 85% of tests is achieved.

Table 12.6 Results of Test Problems

PROBLEM NUMBER	1	2	3	4	5
Real practice	21	17	20	27	23
Exact method	17	10	19	25	21
Proposed algorithm	17	10	19	26	21
PROBLEM NUMBER	6	7	8	9	10
Real practice	20	19	26	14	16
Exact method	17	15	19	14	13
Proposed algorithm	17	15	21	14	13
PROBLEM NUMBER	11	12	13	14	15
Real practice	10	17	28	18	16
Exact method	8	10	20	18	8
Proposed algorithm	8	10	22	18	8
PROBLEM NUMBER	16	17	18	19	20
Real practice	17	8	15	8	12
Exact method	13	7	11	6	12
Proposed algorithm	15	7	11	6	12

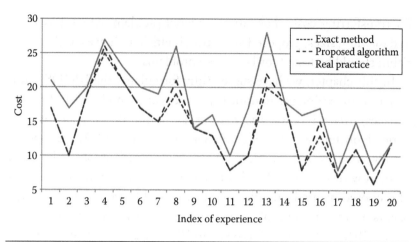

Figure 12.7 Comparison between exact method, proposed algorithm, and real-world case, in terms of total cost.

The exact method gives us the best solutions in all the experiences because our problems are small and medium, but for large problems, this exact method might not be able to answer, so we should try a heuristic algorithm for large-size problems. Our algorithm gives the best answer 80% of times. This proposed algorithm can solve the problems in an average of 0.7 s, while the exact method solves them in an average of 43 s, so it shows that for large-size problems, this reduced solving time is better.

The graphical comparison between the exact method, the proposed algorithm, and real-world case, in terms of total cost, is demonstrated in Figure 12.7. As you see in this figure, the exact method reduces the objective function value in 17 problems (85% of problems), and in 3 problems, total cost is equal in all methods.

12.6 Conclusions

Our method shows that our model might be useful for many companies that hope to reduce their transfer costs and increase the satisfaction of their customers. For large problems, we propose the heuristic algorithm.

Our goal for working on this problem was to decrease hidden costs, which were created by dissatisfaction of customers from late or soon delivery, although we should consider the best and minimum costs for transferring costs. For our considered problem, which has small and medium sizes, proposed algorithm is efficient; however, this algorithm should be tested for large-size problems.

In this chapter, we have successfully formulated the IPODS problem as a mixed-integer nonlinear programming model with different vehicles, time windows, and routing method of distribution. A heuristic algorithm, which is a combination of GA and SA, is proposed to solve the research problem. We solved 20 real problems by exact method with Lingo 11.0 and compared the results with the solutions of proposed heuristic algorithm and real-world cases. Computational results showed that our solution method is effective and efficient. It can reduce 20% of costs, which is equal to 1035 units of costs, and our proposed algorithm can calculate the best solution for 80% of problems.

For the future research, we suggest improving and using the proposed solution method as software for enterprises, which takes data from users and gives the best solution of the sequencing and scheduling problem for each horizon.

References

Averbakh, I., Z. Xue. 2007. On-line supply chain scheduling problems with preemption. *European Journal of Operational Research* 181:500–504.

Ball, M.O., T.L. Magnanti, C.L. Monma, G.L. Nemhauser. 1995. *Network Routing. Handbooks in Operations Research and Management Science*, Vol. 8. North-Holland, Amsterdam, the Netherlands.

Bilgen, B., I. Ozkarahan. 2004. Strategic tactical and operational production-distribution models: A review. *International Journal of Technology Management* 28:151–171.

Chang, Y.C., C.Y. Lee. 2004. Machine scheduling with job delivery coordination. *European Journal of Operational Research* 158:470–487.

Chen, B., C.Y. Lee. 2008. Logistics scheduling with batching and transportation. *European Journal of Operational Research* 189:871–876.

Chen, Z.L. 2004. Integrated production and distribution operations: Taxonomy, models, and review. In D. Simchi-Levi, S.D. Wu, Z.J. Shen (eds.), *Handbook of Quantitative Supply Chain analysis: Modeling in the E-Business Era*. Kluwer Academic Publishers, Norwell, MA.

Chen, Z.L. 2010. Integrated production and outbound distribution scheduling: Review and extensions. *Operation Research* 1:130–148.

Chen, Z.L., G. Pundoor. 2006. Order assignment and scheduling in a supply chain. *Operations Research* 54:555–572.

Chen, Z.L., G. Pundoor. 2009. Integrated order scheduling and packing. *Production Operation Management* 18:672–692.

Chen, Z.L., G.L. Vairaktarakis. 2005. Integrated scheduling of production and distribution operations. *Management Science* 51:614–628.

Cheng, T.C.E., V.S. Gordon, M.Y. Kovalyov. 1996. Single machine scheduling with batch deliveries. *European Journal of Operational Research* 94:277–283.

Dawande, M., H.N. Geismar, N.G. Hall, C. Sriskandarajah. 2006. Supply chain scheduling: Distribution systems. *Production and Operations Management* 15:243–261.

Garcia, J.M., S. Lozano. 2004. Production and delivery scheduling problem with time windows. *Computers and Industrial Engineering* 48:733–742.

Garcia, J.M., S. Lozano. 2005. Production and vehicle scheduling for ready-mix operations. *Computers and Industrial Engineering* 46:803–816.

Garcia, J.M., S. Lozano, D. Canca. 2004. Coordinated scheduling of production and delivery from multiple plants. *Robotics and Computer-Integrated Manufacturing* 20:191–198.

Garcia, J.M., K. Smith, S. Lozano, F. Guerrero. 2002. A comparison of GRASP and an exact method for solving a production and delivery scheduling problem. *Proceedings of Hybrid Information Systems: Advances in Soft Computing*. Physica-Verlag, Heidelberg, Germany, pp. 431–447.

Gharbi, A., M. Haouari. 2002. Minimizing makespan on parallel machines subject to release dates and delivery times. *Journal of Scheduling* 5:329–355.

Goetschalcks, M., C.J. Vidal, K. Dogan. 2002. Modeling and design of global logistics systems: A review of integrated strategic and tactical models and design algorithms. *European Journal of Operational Research* 143:1–18.

Hall, L.A., D.B. Shmoys. 1992. Jackson's rule for single-machine scheduling: Making a good heuristic better. *Mathematics of Operations Research* 17:22–35.

Hall, N.G., C.N. Potts. 2003. Supply chain scheduling: Batching and delivery. *Operations Research* 51:566–584.

He, Y., W. Zhong, H. Gu. 2006. Improved algorithms for two single machine scheduling problems. *Theoretical Computer Science* 363:257–265.

Ji, M., Y. He, T.C.E. Cheng. 2007. Batch delivery scheduling with batch delivery cost on a single machine. *European Journal of Operational Research* 176:745–755.

Kaminsky, P. 2003. The effectiveness of the longest delivery time rule for the flow shop delivery time problem. *Naval Research Logistics* 50:257–272.

Karp, R.M. 1972. Reducibility among combinatorial problem. *Complexity of Computer Computations*. Plenum Press, New York, pp. 85–103.

Lawler, E.L., J.K. Lenstra, A.H.G. Rinnooy Kan, D.B. Shmoys. 1993. Sequencing and scheduling: Algorithms and complexity. *Handbooks in Operations Research and Management Science*, Vol. 4. North-Holland, Amsterdam, the Netherlands, pp. 445–552.

Lee, C.Y., Z.L. Chen. 2001. Machine scheduling with transportation considerations. *Journal of Scheduling* 4:3–24.

Li, C.L., J. Ou. 2005. Machine scheduling with pickup and delivery. *Naval Research Logistics* 52:617–630.

Li, C.L., J. Ou. 2007. Coordinated scheduling of customer orders with decentralized machine locations. *IIE Transactions* 39:899–909.

Li, C.-L., G. Vairaktarakis. 2007. Coordinating production and distribution of jobs with bundling operations. *IIE Transactions* 39:203–215.

Li, C.L., G. Vairaktarakis, C.Y. Lee. 2005. Machine scheduling with deliveries to multiple customer locations. *European Journal of Operational Research* 164:39–51.

Li, K.P., V.K. Ganesan, A.I. Sivakumar. 2005. Synchronized scheduling of assembly and multi-destination air-transportation in a consumer electronics supply chain. *International Journal of Production Research* 43: 2671–2685.

Liu, Z., T.C.E. Cheng. 2002. Scheduling with job release dates, delivery times and preemption penalties. *Informatics Processing Letters* 82:107–111.

Mastrolilli, M. 2003. Efficient approximation schemes for scheduling problems with release dates and delivery times. *Journal of Scheduling* 6:521–531.

Pinedo, M. 2002. *Scheduling Theory, Algorithms, and Systems*, 2nd edn. Prentice Hall, Upper Saddle River, NJ.

Pundoor, G., Z.L. Chen. 2005. Scheduling a production-distribution system to optimize the tradeoff between delivery tardiness and total distribution cost. *Naval Research Logistics* 52:571–589.

Van Buer, M.G., D.L. Woodruff, R.T. Olson. 1999. Solving the medium newspaper production/distribution problem. *European Journal of Operational Research* 115:237–253.

Wang, G., T.C.E. Cheng. 2000. Parallel machine scheduling with batch delivery costs. *International Journal of Production Economics* 68:177–183.

Wang, H., C.Y. Lee. 2005. Production and transport logistics scheduling with two transport mode choices. *Naval Research Logistics* 52:796–809.

Woeginger, G.J. 1994. Heuristics for parallel machine scheduling with delivery times. *Acta Informatica* 31:503–512.

Woeginger, G.J. 1998. A polynomial-time approximation scheme for single-machine sequencing with delivery times and sequence independent batch set-up times. *Journal of Scheduling* 1:79–87.

Zhong, W., G. Dosa, Z. Tan. 2007. On the machine scheduling problem with job delivery coordination. *European Journal of Operational Research* 182:1057–1072.

13

An Integrated Replenishment and Transportation Model

Computational Performance Assessment

RAMEZ KIAN, EMRE BERK, AND ÜLKÜ GÜRLER

Contents

13.1 Introduction

Transformation processes with multiple inputs typically exhibit non-linearities in their output with respect to input usages. They have been traditionally modeled via production functions in the microeconomics literature (Heathfield and Wibe, 1987). One of the most common production functions is the Cobb–Douglas (C–D) production function. This production function assumes that multiple (n) inputs (also called factors or resources) are needed for output, Q, and they may be substituted to take advantage of the marginal cost differentials. In general, it has the form $Q = A \prod_{i=1}^{n} \left[x^{(i)} \right]^{\alpha_i}$, where A represents the total factor productivity of the process given the technology level, $x^{(i)}$ denotes the amount of input i used, and $\alpha i > 0$ is the input elasticity. The total elasticity parameter $r\left(=\dfrac{1}{\displaystyle\sum_{i=1}^{m}\alpha_i}\right)$ may be greater than (smaller than)

or equal to 1 depending on whether there is diminishing (increasing) returns to resources, resulting in convex (concave) operational costs. The C–D production function was first introduced to model the labor and capital substitution effects for the US manufacturing industries in the early twentieth century (Cobb and Douglas, 1928). Despite its macroeconomic origins, since then, it has been widely applied to individual transformation processes at the microeconomic level, as well. For example, the C–D production function was employed to model production processes in the steel and oil industries by Shadbegian and Gray (2005) and in agriculture by Hatirli et al. (2006). Logistics activities associated with shipment preparation, transportation/delivery, and cargo handling also use, directly and/or indirectly, multiple resources such as labor, capital, machinery, materials, energy, and information technology. Therefore, it is not surprising that there is a growing literature on the successful applications of the C–D-type production functions to model the operations in the logistics and supply chain management context. Chang's (1978) work seems to be the earliest to construct a C–D production function to analyze the productivity and capacity expansion options of a seaport. Rekers et al. (1990) estimate a C–D production function for port terminals and specifically model cargo handling service. In a similar vein, Tongzon (1993) and Lightfoot et al. (2012) consider cargo handling processes at container terminals for their production functions. In a recent work, Cheung and Yip (2011) analyze the overall port output via a C–D production function. Studies on technical efficiency in cargo handling and port operations provide additional support for the C–D-type functional relationships, where output is typically measured in volume of traffic (in terms of twenty-foot equivalent unit—TEUs) and inputs may be as diverse as number or net usage time of cranes, types of cranes, number of tug boats, number of workers or gangs, length and surface of the terminals, berth usage, volume carried by land per berth, and energy (e.g., Notteboom et al. 2000, Cullinane 2002, Estache et al. 2002, Cullinane et al. 2002, 2006, Cullinane and Song 2003, 2006, Tongzon and Heng 2005). Comprehensive surveys can be found in

Maria Manuela Gonzalez and Lourdes Trujillo (2009), Trujillo and Diaz (2003), Tovar et al. (2007), and Gonzalez and Trujillo (2009). For land transportation, we may cite the evidence from Williams (1979) and for supply chain management, Ingene and Lusch (1999) and Kogan and Tapiero (2009).

Although multi-input activities in the area of logistics have received the attention of researchers for economic modeling and efficiency measurements, this body of knowledge has been only partially incorporated into decision making at the operational level. As Lee and Fu (2014) observed, the most commonly used transportation cost structures are tapering rates, proportional rates, and blanket rates (Lederer 1994, Taaffe et al. 1996, Ballou 2003, Coyle et al. 2008). Hence, scale economies are the most frequently made assumption. (See also Xu [2013] in a location context.) However, we believe that this assumption ignores the fundamental economic fact that output is typically nonincreasing in the input usage. That is, a C–D production function with total input elasticities being less than unity results in optimal input usage with usage costs being convex in the output level. Our work has been motivated by that the existing literature on the dynamic joint replenishment and transportation models lacks incorporation of the economic production functions. Incorporation of such functions of transportation/delivery activities into the existing logistics management models yields interesting theoretical and practical insights. First, these empirically supported functions, typically, result in the models to be nonlinear and convex in the decision variables for certain parameter settings. For such settings, the theoretical findings of the classical models do not hold any longer. Hence, these new settings are of theoretical interest. Second, the solution methodologies suitable and satisfactory for the classical models become less useful and, in some cases, even unusable. This necessitates the development of novel heuristics. (For a detailed discussion of both aspects in a dynamic lot-sizing framework, see Kian et al. 2014.) In this work, we focus on the suitability of the existing generic solvers and their computational performance for a logistics model with convex costs.

We envision a firm that produces a single product and delivers the production quantity to its vendor-managed inventory warehouse. We consider the dynamic joint replenishment and transportation

274 RAMEZ KIAN ET AL.

problem for this integrated two-stage inventory system where the delivery times of the items from the production site to the warehouse and from the warehouse to a customer's site are negligible, but the logistical operations associated with shipment preparation, transportation/delivery, and cargo handling are nonlinear in the shipment quantity. In particular, we assume that the quantity transported requires multiple inputs whose usage is expressed by a C–D-type production function so that the resulting transportation costs are convex. Therefore, our work differs greatly from the existing models on replenishment and inbound/outbound logistics. Among the significant works in this area, we may cite Lippman (1969), Lee (1989), Pochet and Wolsey (1993), Lee et al. (2003), Jaruphongsa et al. (2005), Berman and Wang (2006), Van Vyve (2007), Hwang (2009), and Hwang (2010). Integrated replenishment and transportation problems have close similarity with the dynamic lot-sizing models in mathematical structure and analytical properties. A dynamic lot-sizing model with convex cost functions of a power form has been studied recently by Kian et al. (2014). It was shown that replenishment is possible even with positive on-hand inventory (contrary to the classical Wagner–Whitin model in Wagner and Whitin [1958]), and thereby, a forward solution algorithm does not exist. In lieu of the optimal solution, heuristics were designed and approximate solutions were investigated. For the related literature and the analytical intricacies of the particular lot-sizing model, we refer the reader to the aforementioned work.

The rest of the chapter is organized as follows. In Section 13.2, we present the assumptions of the model and provide three formulations. In Section 13.3, we provide a numerical study and discuss our findings.

13.2 Model

13.2.1 Assumptions

We consider a single item. The problem is of finite horizon length, T. The demand amount in period t is denoted by $d_t(t = 1,...,T)$. All demands are nonnegative and known, but may be different over the planning horizon. No shortages are allowed. The amount of replenishment (production) in period t is denoted by q_t and is uncapacitated. Replenishment in any period t incurs a fixed cost (of setup) K_t (≥ 0) and unit variable cost, p_t. All units replenished in a period are transported to the warehouse; that is, dispatch quantity in a period is the same as the production quantity. Fixed costs associated with shipments are assumed negligible (or, equivalently may be viewed as subsumed in the fixed replenishment cost under the assumed dispatch policy). Each unit shipped in period t incurs a cost of τ_t. Additionally, the transportation and delivery use m (≥ 1) inputs with unit acquisition cost of input i in period t being $a_t^{(i)}$ for $1 \leq i \leq m$. It is assumed that there are no economies of scale in the acquisition of the inputs and that unit acquisition costs are nonspeculative over the problem horizon. These assumptions dictate that a lot-for-lot acquisition policy is optimal for the inputs needed. (A similar set of assumptions are implicitly made for the ingredients/raw materials needed for the replenishment that involves actual manufacturing.) The input usage for transporting qt units of the item in period t is determined through a stationary C–D function as $q_t = \prod_{i=1}^{m} \left[x_t^{(i)} \right]^{\alpha_i}$ with $\alpha_i \geq 0$ for all i. The stationarity of the function parameters are realistic in that the planning problem considered herein would be of very short term compared to the timeframe required for technological changes that would impact the values of the elasticity and total factor productivity parameters. The inventory on hand at the end of period t at the warehouse is denoted by I_t; each unit of ending inventory in the period is charged a unit holding cost of h_t. Without loss of generality, the initial inventory level, I_0, is assumed to be zero. Given that the short-term nature of the decisions, no discounting is assumed over the horizon although it can easily be incorporated into the model. The objective is to find a joint replenishment and transportation plan that determines the timing and amount of production and delivery (q_t) such that total costs over the horizon are minimized.

Before we proceed with the formulations of the problem, a few remarks are in order about the particulars of our problem setting. (1) In the presence of zero fixed costs of shipment, the assumed dispatch policy is optimal. However, with nonzero fixed costs, it would be suboptimal. This particular fixed cost structure has been studied by Jaruphongsa et al. (2005) with zero unit variable costs. Under nonspeculative (fixed and unit) costs, it has been established that the replenishment quantity in any period k needs to be either zero or equal to the sum of a number of future dispatch quantities. In our setting, we chose fixed shipment costs to be zero for the impact of the special nature of the variable costs to be brought to the foreground. (2) Since Lippman (1969), the shipments have taken into account cargo capacity of individual vehicles and considered stepwise cost structures. Again, for better exposition of the special cost function we assume herein, we ignore this aspect. Thus, our results may be viewed as a relaxation of this cargo capacity constraint. (3) The dynamic lot-sizing problems are special cases of the joint replenishment and transportation problems and, thereby, show close affinity with them under certain cost structures and policies. This is true in our setting, as well. The characteristics of the model herein are similar to those of Kian et al. (2014), and the two-echelon inventory system may be reduced to the single location lot-sizing model studied in the mentioned work. Therefore, in this work, we focus on the computational issues.

13.2.2 Formulations

We first formulate the problem as a mixed-integer nonlinear programming (MINLP) problem. We will consider two equivalent variants. In the first formulation, P_T^1, the decision variables are the replenishment (and shipment) quantities q_t, the binary variables y_t for replenishment setup, the input quantities $x_t^{(i)}$ for $i = 1, \ldots, m$ with the intermediate inventory variables I_t for $1 \leq t \leq T$. The objective function is linear in the variables, but the constraints contain the nonlinear production function that relates the inputs to the replenishment/shipment quantity. In the second formulation, P_T^2, we first determine the optimal input usage for any replenishment/shipment quantity (which may be viewed as preprocessing) and incorporate the production function relationship into the objective function rendering the problem into a form with a nonlinear objective function with only linear constraints. In P_T^2, the decision variables are the replenishment (and shipment) quantities q_t, the binary variables y_t for replenishment setup with the intermediate inventory variables I_t for $1 \leq t \leq T$.

We state the first formulation P_T^1, which acts as a building block for the second formulation, formally as follows:

$$\min \sum_{t=1}^{T} \left[K_t y_y + \left(p_t + \tau_t \right) q_t + \sum_{i=1}^{m} \left(a_t^{(i)} x_t^{(i)} \right) + h_t I_t \right] \text{ s.t.} \qquad (13.1a)$$

$$M y_t \geq q_t \quad t \in \{1,\ldots,T\} \qquad (13.1b)$$

$$I_t = I_{t-1} + q_t - d_t \quad t \in \{1,\ldots,T\} \qquad (13.1c)$$

$$q_t = A \prod_{i=1}^{m} \left[x_t^{(i)} \right]^{\alpha_i} \quad t \in \{1,\ldots,T\} \qquad (13.1d)$$

$$y_t \in \{0,1\}, x_t^{(i)} \geq 0, \quad q_t \geq 0, \quad i \in \{1,\ldots,m\}, t \in \{1,\ldots,T\} \qquad (13.1e)$$

where M is a sufficiently large positive number. The first set of constraints (13.1b) ensures that setups are performed only in the periods in which replenishment is positive, (13.1c) gives the evolution of on-hand inventories, (13.1d) represents the production function relating the inputs and the transported quantity, and (13.1e) are binary and nonnegativity constraints. We assume that the initial inventory is zero and these demands are net demands. The second formulation P_T^2 is obtained from P_T^1 by first deriving the optimal input allocations for a given shipment quantity. To this end, consider the subproblem where the input acquisition costs in period t are minimized given $q_t = Q$. As the input usage is uncapacitated, the first-order conditions imply that, for any i and j, $j \in \{1,\ldots,m\}$,

$$x_t^{(i)}(Q)^* = \frac{\alpha_i a_t^{(j)}}{\alpha_j a_t^{(i)}} x_t^{(j)}(Q)^* \qquad (13.2)$$

where $x_t^{(i)}(Q)^*$ is the optimal usage of input i to transport Q units of the item. Hence, for $1 \leq i \leq m$,

$$x_t^{(i)}(Q)^* = \frac{\alpha_i}{a_t^{(i)}} A^{-r} \prod_{j=1}^{m} \left(\frac{a_t^{(j)}}{\alpha_j} \right) \alpha_j^r Q^r \qquad (13.3)$$

(For details, see Heathfield and Wibe 1987.) Correspondingly, for a shipment quantity Q, the minimum transportation cost in period t, $C_t^*(Q)$, becomes

$$C_t^*(Q) = w_t Q^r + \tau_t Q \qquad (13.4)$$

where

$$w_t = \left(\frac{1}{r}\right) A^{-r} \prod_{j=1}^{m} \left(\frac{a_t^{(j)}}{\alpha_j}\right)^{\alpha_j r}$$

The expression of $C^*(Q)$ enables us to rewrite the MINLP formulation as P_T^2 as follows:

$$\min \sum_{t=1}^{T} \left[K_t y_t + p_t q_t + C_t^*(q_t) + h_t I_t \right] \; s.t. \qquad (13.5a)$$

$$M y_t \geq q_t \quad t \in \{1,\ldots,T\} \qquad (13.5b)$$

$$I_t = I_{t-1} + q_t - d_t \quad t \in \{1,\ldots,T\} \qquad (13.5c)$$

$$y_t \in \{0,1\}, \quad q_t \geq 0, \quad i \in \{1,\ldots,m\}, \quad t \in \{1,\ldots,T\} \qquad (13.5d)$$

where M is as defined before. The constraints (13.5b), (13.5c), and (13.5d) perform the same function as in P_T^1, but we have been able to eliminate the input variables and to render all constraints linear at the expense of nonlinearizing the objective function. Clearly, the second formulation is more compact and has computational advantages as demonstrated in our numerical study. We can also formulate the problem as a dynamic programming (DP) problem. Define $J_t^T(I_t)$ as the minimum total cost under an optimal joint replenishment and transportation plan for periods t through T, where I_t is the ending inventory as defined before in the recursions (13.1c) or (13.5c). Then,

$$J_{t-1}^T(I_{t-1}) = \min_{q_t \geq \max(0, d_t - I_{t-1})} \left\{ K_t 1_{\{q_t > 0\}} + h_t I_t + p_t q_t + C_t^*(q_t) + J_t^T(I_t) \right\}$$

$$t \in \{1,\ldots,T\} \qquad (13.6)$$

where $1_{\{q_t > 0\}}$ indicates the existence of a setup in period t, with the boundary condition in period T being $J_T^T(I_T) = 0$ for any $I_T \geq 0$. The optimal solution is found using the earlier recursion, and $J_0^T(0)$ denotes the minimum cost over the problem horizon.

The main difficulty with this formulation is its high dimensionality. The memory requirements and the system state size become prohibitively large, and the solution times are too long. It is not suitable for problems of large sizes in terms of horizon lengths and/or demand values. For our work, this formulation is important in that it provides a guaranteed optimal solution and serves as the benchmark in our numerical study.

13.3 Numerical Study

For our numerical study, we constructed our experiment set in line with Kian et al. (2014).

We considered a problem horizon of $T = 100$ periods. Period demands are generated randomly from three normal distributions with respective coefficients of variation, $cov = 0.8$, 0.4, and 0.2 and standard deviation σ ($= 40$) where negative demand values have been replaced with zero demands. We denote the three demand patterns by $D1$, $D2$, and $D3$, respectively. All other system parameters are stationary. Noting that unit replenishment cost p_t and unit transportation cost τ_t can be subsumed into h_t by simple transformations through inventory recursions, we assume them to be negligible over the entire problem horizon. We set unit holding cost rate, $h_t = h = 1$, and setup cost is selected as a function of the mean demand rate, $K_t = K = [J^2/2]\mu$, where J may be viewed as a proxy for the average size of a replenishment quantity under the simple EOQ formula. We have $J \in \{2, 3, 4, 5\}$. We considered $r = 1.5$. This corresponds to the C–D-type economic production function with convex costs. To select the parameters for the nonlinear transportation/delivery component, we used the formulation P_2^T as the base. For this formulation, we set $w_t = w$ and considered the variable cost of transportation per unit when a dispatched quantity equals the average demand per period, $\bar{w}$ where $w = [w\mu^r]/\mu = w\mu^{r-1}$. Letting $a = h/\bar{w}$, we have $w = h\mu/(a\mu^r)$ with $a \in \{0.02, 0.05, 0.1\}$ so that the resulting variable cost for a shipment quantity of q units is given by $[h\mu/a](q/\mu)r$. Note that $\bar{w}$ is decreasing in a. The same sets of 10 demand realizations generated for each demand distribution were used for all experiment instances throughout the study. Overall, we have $120 = (4 \times 3 \times 10)$ experiment instances for P_T^2. As part of our study, we also tested the efficacy of formulation P_T^2, which is structurally different from P_T^2.

For consistency, we selected the parameters for this formulation as follows. We considered three values of number of iso-elastic inputs, $m = 1, 2, 5$ and $\alpha_i = \alpha$ for $1 \le i \le m$ with $m\alpha = 1/r$. (All other parameters were selected as for P_T^2.) Overall, we have $360 = (3 \times 4 \times 3 \times 10)$ experiment instances for P_T^2. The optimal plan has been obtained by the DP algorithm discussed earlier. We tested the solvers AlphaECP, Baron, Bonmin, Couenne, LINDOGlobal, and KNITRO available online at the NEOS server (http://www.neos-server.org/neos/solvers/index. html). The server's goal has been described as specifying and solving optimization problems with minimal user input (Dolan et al. 2002). The solver defaults/options were set at their defaults except that the time limits on all have been set to 1500 s since lower time resources resulted in too many interrupts in preliminary tests.

In our numerical study, (1) we considered an overall assessment of the computational performances of the two formulations with respect to the demand patterns and the number of inputs using different optimizers, and (2) focusing on the formulation P_T^2, we used the ANalysis Of VAriance (ANOVA) to identify the factors that have statistically significant impact on the solution quality.

13.3.1 Overall Assessment

The performance measures are (1) the number of instances in which a feasible solution has been obtained by a solver, and (2) the percentage deviation from the optimal solution for the obtained solutions averaged over all 120 experiment instances for a particular demand distribution. Note that in the latter computation, the experiment instances in which a solver failed have been excluded.

We begin our analysis with our findings on formulation P_T^1. The overall performance summary with $m = 1, 2, 5$ for the entire experiment set for this formulation is presented in Table 13.1, where # denotes the first performance measure and % denotes the second. For the cases when no feasible solution was obtained, an m-dash (—) has been used to denote the unavailable second measure.

AlphaECP failed to obtain a solution in all experiment instances, whereas LINDOGlobal was able to obtain a solution in all experiment instances except for the demand distribution $D2$. However, for that pattern, it also resulted in a solution in the most number of instances.

Table 13.1 Overall Summary of Performance Measures

		ALPHAECP		BARON		BONMIN		COUENNE		LINDOGLOBAL		KNITRO	
		%	#	%	#	%	#	%	#	%	#	%	#
P_T^1 $m=1$	D1	—	0	119.37	118	0.00	30	63.65	96	3.51	120	56.25	19
	D2	—	0	161.44	82	0.00	27	75.35	72	1.26	108	5.70	38
	D3	—	0	159.94	120	0.00	30	73.91	120	0.82	120	9.29	45
P_T^1 $m=2$	D1	—	0	—	0	0.00	27	86.91	72	18.74	120	6.48	87
	D2	—	0	—	0	0.00	30	—	0	15.05	72	1.43	53
	D3	—	0	—	0	0.00	27	94.91	48	11.67	120	1.14	48
P_T^1 $m=5$	D1	—	0	—	0	0.00	27	71.62	84	231.66	120	23.81	52
	D2	—	0	—	0	0.00	30	79.96	60	199.35	72	12.92	32
	D3	—	0	—	0			80.16	120	272.1	120	6.16	106
P_T^2	D1	1.45	120	15.57	120	2.90	120	1.28	120	5.44	120	6.20	120
	D2	0.53	120	4.03	120	0.86	120	0.67	120	1.54	120	1.23	120
	D3	0.38	120	1.36	120	0.53	120	0.45	120	1.02	120	0.81	120

Bonmin has low performance in obtaining a solution, but the quality of the obtained solution is very good (optimal in many instances). Regardless of the number of inputs in the system, it was able to get a near-optimal solution for $D1$. The distribution $D2$ seems to present the most difficulty for given m and other parameters except for Bonmin.

For LINDOGlobal, the number of inputs in the problem setting has a negative impact on the quality of the obtained solutions. For other solvers, the behavior may not be monotone (e.g., KNITRO, Bonmin). However, in a very general qualitative sense, we get the impression that solver performance (in both criteria) tends to worsen as the number of inputs increases in the problem setting. This observation has motivated us to construct the second formulation, P_T^2. For P_T^2, the performances of all solvers have improved significantly in terms of the number of instances for which a feasible solution was obtained; none of the solvers failed across the entire experimental bed. Also, the solution quality for all solvers except LINDOGlobal (for $m=1$ case) has increased. These indicate that the formulation P_T^2 is more amenable to use on the available solvers.

13.3.2 ANOVA Assessment

The overall assessment presented earlier was based on the performances of the two formulations and the solvers in an aggregate sense. Next, we focus on the formulation P_T^2 and use the formal statistical tool ANOVA to identify the factors that impact the solution quality significantly in a statistical sense.

We considered a three-way ANOVA where the factors are (1) K (representing the fixed replenishment cost) considered in four levels K_i, $i=1,\ldots, 4$; (2) W (representing the transportation cost coefficient, w) considered in three levels, $W_j, j = 1, 2, 3$ as given earlier in the experimental bed; and (3) the different solvers denoted by S with six levels, S_k, $k = 1,\ldots,6$ corresponding to the solvers in the order given earlier with $n=10$ replications (corresponding to the demand realizations) at each experimental instance. The response variables y_{ijkl}, $i = 1,\ldots,4$; $j = 1,2,3$; $k = 1,\ldots,6$; and $l = 1,\ldots,10$ are taken as the percentage deviations of the solutions provided by the solvers from the optimal solution, which is obtained by DP. The ANOVA study was conducted for each demand distribution separately.

Table 13.2 ANOVA for $D1$

SOURCE	DF	SEQ SS	ADJ SS	ADJ MS	F	P
K	3	18285.12	18285.12	6095.04	1246.5	0
W	2	4611.81	4611.81	2305.9	471.58	0
S	5	17151.57	17151.57	3430.31	701.54	0
$K*W$	6	12221.28	12221.28	2036.88	416.56	0
$K*S$	15	7259.13	7259.13	483.94	98.97	0
$W*S$	10	2259.74	2259.74	225.97	46.21	0
$K*W*S$	30	6201.09	6201.09	206.7	42.27	0
Error	648	3168.54	3168.54	4.89		
Total	719	71158.26				

The ANOVA tables for the three distributions are given in Tables 13.2 through 13.4. The performance statistics for each factor level computed across the other experiment parameters are tabulated in Table 13.5 for each demand distribution. Finally, the interaction effects of the factor levels are provided in Figures 13.1 through 13.3 for the each distribution, respectively. The inspection of these results reveals the following findings.

Firstly, all the factors and the interactions have significant impact on the solution quality, which is indicated by very large F values and correspondingly very small P-values, implying that the hypothesis that states that all factor levels have the same effect on the response variable is rejected for all three distributions. A closer inspection of the results provides further information regarding (1) the relative impact of the factors, (2) direction of the factor-level impact, and (3) the interaction effect. We treat each demand distribution separately.

Table 13.3 ANOVA for $D2$

SOURCE	DF	SEQ SS	ADJ SS	ADJ MS	F	P
K	3	640.025	640.025	213.342	754.86	0
W	2	542.101	542.101	271.051	959.05	0
S	5	1020.997	1020.997	204.199	722.52	0
$K*W$	6	1163.260	1163.26	193.877	685.99	0
$K*S$	15	134.743	134.743	8.983	31.78	0
$W*S$	10	120.064	120.064	12.006	42.48	0
$K*W*S$	30	347.503	347.503	11.583	40.99	0
Error	648	183.140	183.14	0.283		
Total	719	4151.832				

Table 13.4 ANOVA for $D3$

SOURCE	DF	SEQ SS	ADJ SS	ADJ MS	F	P
K	3	451.116	451.116	150.372	3136.61	0
W	2	353.541	353.541	176.771	3687.26	0
S	5	86.901	86.901	17.38	362.53	0
$K*W$	6	855.188	855.188	142.531	2973.06	0
$K*S$	15	82.740	82.74	5.516	115.06	0
$W*S$	10	63.062	63.062	6.306	131.54	0
$K*W*S$	30	218.256	218.256	7.275	151.75	0
Error	648	31.066	31.066	0.048		
Total	719	2141.870				

Consider Table 13.2. Comparing the F values, we observe that the most important factors are, respectively, K, S, W, and the two-way KW interaction. From Table 13.5, we see that $K4$, $W3$, and $S2$ (Solver Baron) result in the worst solution quality on average. Next, inspecting the impact of average effect of different levels of factors from Table 13.5, we see that the largest deviation from the optimal results is observed when fixed cost is highest at $K4$ level, when W is at the $W3$ level, and the solver $S2$ is used. From Figure 13.1, we observe that the differential effect as K increases depends on the level of W implying a significant interaction of K and W with the worst performance occurring at $K4W3$ combination. Although not as significant, there is also some interaction of K with the solvers. As K level changes from 3 to 4, the performance deteriorates significantly with solvers $S5$ (LINDOGlobal) and $S6$ (KNITRO). A similar relation also holds regarding the interaction between W and the solvers.

Similar analysis for $D2$ and $D3$ reveals the following. For $D2$, the factors with the highest F values are ordered as W, K, S, and the two-way interaction KW. Table 13.5 shows that there are less drastic differences between the average solution quality corresponding to different levels of the factors. Figure 13.2 shows that the KW interaction is still significant, and the difference between the levels of K is highest for $W3$, where the interaction of solvers with K and W is reduced. The ordering of solver performances is similar to that of $D1$. For $D3$, we note that the factors with the highest F values are ordered as W, K, KW, and S. We again observe that the average solution quality corresponding to different factor levels generally becomes closer

Table 13.5 Response Variable Statistics for Different Factors

		MIN	MAX	AVE.		MIN	MAX	AVE.		MIN	MAX	AVE.
D1	K1	0	18.753	2.354	W1	0	18.753	2.979	S1	0	11.85	1.451
	K2	0	17.793	2.205	W2	0	33.234	4.578	S2	5.931	53.504	15.574
	K3	0	16.648	3.16	W3	0	60.361	8.87	S3	0	19.52	2.897
	K4	0.547	60.361	14.181					S4	0	6.934	1.283
									S5	0	50.718	5.438
									S6	0	60.361	6.204
D2	K1	0	5.44	0.878	W1	0	5.441	1.61	S1	0	4.088	0.531
	K2	0	5.167	0.656	W2	0	4.118	0.662	S2	1.4	14.351	4.029
	K3	0	6.166	1.316	W3	0	2.867	0.362	S3	0	6.423	0.858
	K4	0	14.35	3.057					S4	0	3.753	0.669
									S5	0	12.708	1.542
									S6	0	12.708	1.23
D3	K1	0	1.4	0.236	W1	0	1.86	0.28	S1	0	3.48	0.37
	K2	0	1.86	0.201	W2	0	1.56	0.244	S2	0.141	9.099	1.359
	K3	0	2.053	0.476	W3	0	9.09	1.748	S3	0	4.595	0.527
	K4	0	9.099	2.11					S4	0	3.024	0.45
									S5	0	8.848	1.017
									S6	0	8.848	0.811

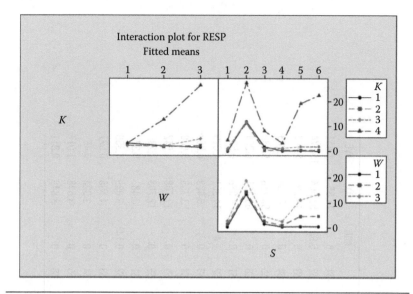

Figure 13.1 Factor interaction for *D*1.

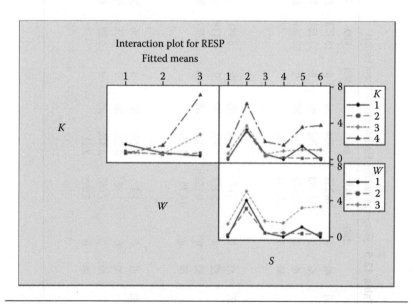

Figure 13.2 Factor interaction for *D*2.

to each other, while the *KW* interaction is still emphasized and the interactions with the solvers become less emphasized.

From the earlier analysis, we see that the solvers' performances get more and more closer to each other as the coefficient of variation of the demand distribution gets smaller and the worst performances

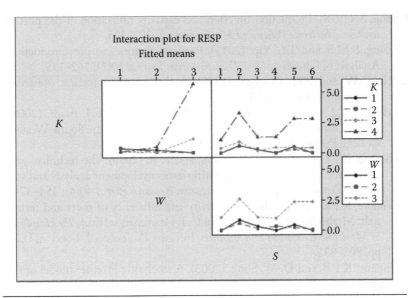

Figure 13.3 Factor interaction for $D3$.

are observed for the $K4W3$ large fixed cost and low transportation cost coefficient combination. Furthermore, $S1$, $S3$, and $S4$ (Solvers AlphaECP, Bonmin, and Couenne, respectively) are always among the best three performing solvers (although their ordering may change), whereas the worst performer is $S2$ in all three demand distributions. We observe that solver performances depend drastically on problem formulations as well as cost parameters. We should also mention that they may as well depend on possible user interventions such as initial point selections that were not imposed in our study.

Acknowledgment

The work of Ramez Kian is partially supported by TUBİTAK (The Scientific and Technological Research Council of Turkey).

References

Ballou, R.H. (2003). *Business Logistics/Supply Chain Management*, 5th edn. Prentice Hall, Upper Saddle River, NJ.

Berman, O. and Q., Wang. (2006). Inbound logistic planning: Minimizing transportation and inventory cost. *Transportation Science* 40(3): 287–299.

Chang, S. (1978). Production function and capacity utilization of the port of mobile. *Maritime Policy and Management* 5: 297–305.

Cheung, S.M.S. and T.L. Yip. (2011). Port city factors and port production: Analysis of Chinese ports. *Transportation Journal* 50(2): 162–175.

Cobb, C.W. and P.H. Douglas. (1928). A theory of production. *American Economic Review* 8(1): 139–165.

Coyle, J.J., C.J. Langley, B.J. Gibson, R.A. Novack, and E.J. Bardi. (2008). *Supply Chain Management: A Logistics Perspective*, 8th edn. South-Western College Publication, Cincinnati, OH.

Cullinane, K., T.-F. Wang, D.-W. Song, and P. Ji. (2006). The technical efficiency of container ports: Comparing data envelopment analysis and stochastic Frontier analysis. *Transportation Research Part A* 40(4): 354–374.

Cullinane, K.P.B. (2002). The productivity and efficiency of ports and terminals: Methods and applications. In C.T. Grammenos (Ed.), *The Handbook of Maritime Economics and Business*, Informa Professional, London, U.K., pp. 803–831.

Cullinane, K.P.B. and D.-W. Song. (2003). A stochastic Frontier model of the productive efficiency of Korean container terminals. *Applied Economics* 35: 251–267.

Cullinane, K.P.B. and D.-W. Song. (2006). Estimating the relative efficiency of European container ports: A stochastic Frontier analysis. In K.P.B. Cullinane and W.K. Talley (Eds.), *Port Economics, Research in Transportation Economics*, Vol. XVI. Elsevier, Amsterdam, the Netherlands, pp. 85–115.

Cullinane, K.P.B., D.-W. Song, and R. Gray. (2002). A stochastic frontier model of the efficiency of major container terminals in Asia: Assessing the influence of administrative and ownership structures. *Transportation Research A: Policy and Practice* 36: 743–762.

Dolan, E., R. Fourer, J.J. Mor, and T.S. Munson. (2002). Optimization on the NEOS server. *SIAM News* 35(6), 1–5.

Douglas, P.H. (1976). The Cobb–Douglas production function once again: Its history, its testing, and some new empirical values. *Journal of Political Economy* 84(5): 903–916.

Estache, A., M. Gonzalez, and L. Trujillo. (2002). Efficiency gains from port reform and the potential for yardstick competition: Lessons from Mexico. *World Development* 30(4): 545–560.

Gonzalez, M.M. and L. Trujillo. (2009). Efficiency measurement in the port industry: A survey of the empirical evidence. *Journal of Transport Economics and Policy* 43(Part 2): 157–192.

Hatirli, S.A., B. Ozkan, and C. Fert. (2006). Energy inputs and crop yield relationship in greenhouse tomato production. *Renewable Energy* 31(4): 427–438.

Heathfield, D. and S. Wibe. (1987). *An Introduction to Cost and Production Functions.* Humanities Press International, New Jersey, NJ.

Hwang, H.C. (2009). Inventory replenishment and inbound shipment scheduling under a minimum replenishment policy. *Transportation Science* 43(2): 244–264.

Hwang, H.C. (2010). Economic lot-sizing for integrated production and transportation. *Operations Research* 58(2): 428–444.

Ingene, C.A. and R.F. Lusch. (1999). Estimation of a department store production function. *International Journal of Physical Distribution & Logistics Management* 29(7/8): 453–464.

Jaruphongsa, W., S. Cetinkaya, and C.-Y. Lee. (2005). A dynamic lot sizing model with multi-mode replenishments: Polynomial algorithms for special cases with dual and multiple modes. *IIE Transactions* 37: 453–467.

Kian, R., Ü. Gürler, and E. Berk. (2014). The dynamic lot-sizing problem with convex economic production costs and setups. *International Journal of Production Economics* 155: 361–379.

Kogan, K. and C.S. Tapiero. (2009). Optimal co-investment in supply chain infrastructure. *European Journal of Operational Research* 192(1): 265–276.

Lederer, P.J. (1994). Competitive delivered pricing and production. *Regional Science and Urban Economics* 24(2): 229–252.

Lee, C.-Y. (1989). A solution to the multiple set-up problem with dynamic demand. *IIE Transactions* 21: 266–270.

Lee, C.-Y., S. Cetinkaya, and W. Jaruphongsa. (2003). A dynamic model for inventory lot sizing and outbound shipment scheduling at a third-party warehouse. *Operations Research* 51(5): 735–747.

Lee, S.-D. and Y.-C. Fu. (2014). Joint production and delivery lot sizing for a make-to-order producer-buyer supply chain with transportation cost. *Transportation Research Part E* 66: 23–35.

Lightfoot, A., G. Lubulwa, and A. Malarz. (2012). An analysis of container handling at Australian ports, *35th ATRF Conference 2012*, Perth, Western Australia, Australia.

Lippman, S.A. (1969). Optimal inventory policy with multiple set-up costs. *Management Science* 16: 118–138.

Notteboom, T.E., C. Coeck, and J. Van den Broeck. (2000). Measuring and explaining relative efficiency of container terminals by means of Bayesian stochastic Frontier models. *International Journal of Maritime Economics* 2(2): 83–106.

Pochet, Y. and L.A. Wolsey. (1993). Lot-sizing with constant batches: Formulations and valid inequalities. *Mathematics of Operations Research* 18(4): 767–785.

Rekers, R.A., D. Connell, and D.I. Ross. (1990). The development of a production function for a container terminal in the port of Melbourne. *Papers of the Australiasian Transport Research Forum* 15: 209–218.

Shadbegian, R.J. and W.B. Gray. (2005). Pollution abatement expenditures and plant level productivity: A production function approach. *Ecological Economics* 54(2): 196–208.

Taaffe, E.J., H.L. Gauthier, and M.E. OKelly. (1996). *Geography of Transportation*, 2nd edn. Prentice-Hall, Inc., Upper Saddle River, NJ.

Tongzon, J. and W. Heng. (2005). Port privatization, efficiency and competitiveness: Some empirical evidence from container ports (Terminals). *Transportation Research Part A* 39: 405–424.

Tongzon, J.L. (1993). The Port of Melbourne Authority's pricing policy: Its efficiency and distribution implications. *Maritime Policy and Management* 20(3): 197–203.

Tovar, B., S. Jara-Daz, and L. Trujillo. (2007). Econometric estimation of scale and scope economies within the Port Sector: A review. *Maritime Policy & Management* 34(3): 203–223.

Trujillo, L. and S. Jara-Daz. (2003). *Production and Cost Functions and their Application to the Port Sector: A Literature Survey*, Vol. 3123. World Bank Publications, http://dx.doi.org/10.1596/1813-9450-3123.

Van Vyve, M. (2007). Algorithms for single-item lot-sizing problems with constant batch size. *Mathematics of Operations Research* 32: 594–613.

Wagner, H.M. and T.M. Whitin. (1958). Dynamic version of the economic lot size model. *Management Science* 5(1): 89–96.

Williams, M. (1979). Firm size and operating costs in urban bus transportation. *The Journal of Industrial Economics* 28(2): 209–218.

Xu, S. (2013). Transport economies of scale and firm location. *Mathematical Social Sciences* 66: 337–345.

Author Index

Subject Index

A

Alternate location allocation (ALA), 240, 244, 248

Analysis of variance (ANOVA), 280
distributions, 283–284, 286–287
three-way factor, 282
variable statistics, 283–285

Analytical hierarchy process (AHP), 107, 112–113, 118, 219

Ant colony optimization (ACO) algorithm, 36

Applied VND algorithm, 45–46

ArcGIS, 202

Arc routing for connectivity problem (ARCP)
class partitions, 168
computational complexity (*see* Computational complexity)
computational performance
depot location effects, 190–192
unblocking time case, 188, 190
unblocking time cases, 188, 190

connected graph, 167–168
data acquisition
blocked roads, 180, 189–190
depot points, 179
disconnected components, 180, 189
high-risk roads, 180, 188
Istanbul map, 179–180
Istanbul road representation, 179, 181
low-risk roads, 180, 189
road distances, 179, 182–187
debris management, 171
definition, 167
dynamic path-based model, 171
high-magnitude earthquake, 166
MIP model
constraints, 176–179
decision variables, 175
flow variables, 174, 177
objective function, 176
properties, 174
sets, indices, and input parameters, 175
vehicle balance equations, 176

For Product Safety Concerns and Information please contact our
EU representative GPSR@taylorandfrancis.com Taylor & Francis
Verlag GmbH, Kaufingerstraße 24, 80331 München, Germany